Transparent Metals

David J. Fisher

Published by **Materials Research Forum LLC**
Millersville, PA 17551, USA

Published as part of the book series
Materials Research Foundations
Volume 174 (2025)
ISSN 2471-8890 (Print)
ISSN 2471-8904 (Online)

Print ISBN 978-1-64490-346-9
ePDF ISBN 978-1-64490-347-6

Distributed worldwide by

Materials Research Forum LLC
105 Springdale Lane
Millersville, PA 17551
USA
http://www.mrforum.com

Printed in the United States of America
10 9 8 7 6 5 4 3 2 1

Table of Contents

Introduction 1

Techniques...30

Nano-Material Preparation 55

Applications...85

   Solar Cells 85

   Organic solar cells...88

   Perovskite solar cells..89

   Other Photovoltaics..92

   Heaters...99

   Diffraction Grating..99

   Antennae..99

   Supercapacitors 101

About the Author..102

References...103

Introduction

The title of this work might at first be thought to be an oxymoron, or even an outright error, but this century has seen the expansion of applications for optical materials based upon photonic band-gap structures which are created by combining films of metals and dielectrics. This offers the possibility of producing high transmittance levels in the 'windows' of devices, even when they include a relatively thick film of metal. Such materials have therefore been termed 'transparent metals'. The present materials are used to produce devices having electrically variable optical properties. Such photonic devices exhibit increased speeds of operation, reduced size and increased immunity to temperature changes. The structures can be made from semiconductors, dielectrics and their combinations. A silver|silica multilayer thin-film bandpass filter for the ultraviolet range was, for example, created[1] by using the transparent metals. In order to calculate the transmission of the designed metal-dielectric filters, use was made of the transfer matrix method for lossy structures. The suppression of unwanted visible and infrared parts of the spectrum was possible by using just a small number of layer-pairs. A numerical investigation was made[2] of the properties of metal-dielectric 1-dimensional photonic band-gap structures. Theory predicted that interference effects would give rise to a transparent metallic structure which allowed the transmission of light over a tunable range of frequencies, such as the ultraviolet, the visible or the infrared. The structure could be tailored to block ultraviolet, transmit in the visible range and reflect all other waves of lower frequency, from infrared to microwave and beyond. The transparent metallic structure consisted of a stack of alternating layers of metal and dielectric material, such that its complex index of refraction alternated between a high value and a low value. The structure remained transparent even when the total amount of metal was increased to hundreds of skin-depths in thickness. One-dimensional photonic band-gap structures on plastic substrates have meanwhile long been proposed for electromagnetic shielding applications in the radio frequency range[3].

glass	1mm
LiF	75nm
Ag	10nm
LiF	150nm
Ag	20nm
LiF	50nm
air-gap	variable
LiF	50nm
Ag	20nm
LiF	150nm
Ag	10nm
LiF	75nm
glass	1mm

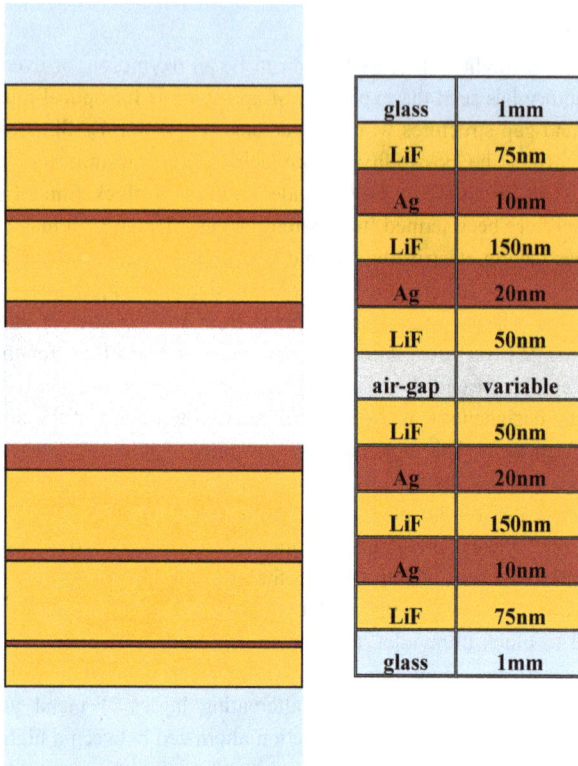

Figure 1. Possible photonic band-gap structure, with 'transparent metals' and controllable air-gap

Transparent electrodes attract intense interest in many technological fields, including opto-electronic devices, transparent film heaters and electromagnetic applications. Most of these 'transparent-metal' windows are much less exotic than in the above case and may be just a very thin metal film, a mere mesh of otherwise-opaque metal or a non-metallic material which happens to possess a conductivity which is worthy of a metal. New-generation transparent electrodes are expected to possess 3 main physical properties: high electrical conductivity, high transparency and mechanical flexibility. The most efficient and widely used transparent conducting material is currently indium tin oxide. The scarcity of indium, together with the oxide's lack of flexibility and the relatively high manufacturing costs, have however prompted a search for alternative materials. The requirement of high optical transparency and high electrical conductivity imposes strict limits on electrodes which are based upon metallic materials. The visible transmittance of the electrodes tends to be lower than the transparent substrates upon which the electrodes are built, and the transmittance relative to the substrate is less than 100%. The properties of dielectric|metal|dielectric structures depend[4] upon the thicknesses of the various parts. The threshold thickness of the metal film is usually about 10nm, whereupon the structures change from insulating to highly conductive. This is attributed to the percolation of conducting metal paths. The transmittance of the films increases when the metal thickness increases up to the percolation thickness. A further increase leads to a decrease in transmittance. A flexible dielectric|metal|dielectric electrode was created[5] which offered about 88.4% absolute transmittance. This was higher than the roughly 88.1% transmittance of the polymer substrate, resulting in a relative transmittance of about 100.3%. This was achieved by designing an optimized dielectric-metal-dielectric structure, and was attributed to an ultra-thin ultra-smooth copper-doped silver film of low optical loss and low sheet-resistance. The metal film had a thickness of about 6.5nm, a roughness of less than 1nm and a sheet-resistance of about $18.6\Omega/sq$.

In addition to metals which are truly optically transparent, if only when in very-thin film-form, there are also metallic materials which are only 'structurally transparent' and consist of a mesh or network arrangement of opaque metal wires. The 'transparent' description is thus used here in an ironic manner, and should be seen in the same light in which 'room-temperature superconductor' is applied to materials which are certainly not (for the moment) really superconducting under ambient conditions. A major advantage of this approach is that it retains optical transparency while minimizing any loss in electrical properties. A mesh structure has the advantage of being able to balance sheet-resistance and optical transparency. These two factors can be modified by choosing mesh parameters such as the width, pitch, thickness and geometry of the pattern. The resistance depends upon the geometrical design of the pattern. In the case of a square grid, for example, the resistance is calculated on the basis of the height (h), width (w) and pitch (p) of the mesh while the transparency is deduced from these mesh parameters under the assumption that the geometrical design is a grid. The relationship between sheet resistance and transparency is then,

$$\text{sheet-resistance} = (\rho/h)[(1 - \sqrt{T} + T)/(1 - \sqrt{T})]$$

where ρ is the resistivity and T is the transparency. The wider the spacing between wires, the greater is the transparency. A coating of conductive polymer or oxide may be needed in order to avoid disadvantages arising from electrochemical impedance. Introducing indium tin oxide into the spaces of a grid greatly increases charge transfer and interfacial area without impairing the transparency. In hybrid micro-electrodes which are based upon indium tin oxide, the metallic grid bridges gaps in the oxide and leads to a very flexible behaviour. The metal also lowers the mechanical load which is imposed on the oxide. Another approach is to use metallic nano-wire structures. These offer the advantages, over mesh, of a high effective surface area, electrical conductivity and optical transparency. With their outstanding physical properties, metallic nanowire-based percolating networks remain one of the most promising alternatives to indium tin oxide[6]. They also have several other advantages, such as solution-based processing, and are compatible with large-area deposition techniques. Costs tend to be lower because of the small quantities of nano-material which are required in order to satisfy industrial performance criteria. In order to facilitate the use of metallic nano-wire networks as a replacement for indium tin oxide, it has been necessary to overcome their inherent stability issues while retaining their properties and cost-effectiveness[7]. Various strategies have been used to model them computationally, from the nano- to the macro-scale, in order to study dynamic failures and identify the mechanisms that account for the failure modes. One proposed method for improving the stability of metallic nano-wire networks has been encapsulation with regard to applications such as solar cells, transparent film heaters and sensors.

Susceptibility to corrosion, and thermal instability, remain limiting factors for their widespread use. One scalable and economical process was suggested[8] which involved using electrophoretic deposition to create a very stable hybrid transparent electrode with a sandwich structure in which a silver nano-wire network was to be covered with a graphene oxide film on each side. The process allowed the conductive transparent film to be transferred to any surface following deposition while retaining a sheet-resistance of 15Ω/sq and a tunable transmittance of 70 to 87% at 550nm. Unlike bare silver nano-wire networks, the hybrid material could retain its original conductivity during long-term storage in up to 80% relative humidity. This was attributed to the absence of silver corrosion-products within the encapsulant. Voltage-ramping and resistance measurements at up to 20V revealed a stabilization mechanism, due to the encapsulant, which delayed failure onset and prevented any sudden deviation of the resistance into the mega-ohm range, which occurred in the case of bare networks. The problem of a relatively high electrochemical impedance also arises because the electron paths are quite limited. Plating gold onto silver nano-wires can lead to a higher stretchability and to more stable electrical properties while retaining an 83% transmittance, with an electrochemical impedance of 1.1 to 3.2Ω/sq. Thin films of indium tin oxide are the predominant transparent electrodes but suffer from brittleness and low infrared transmittance[9]. Alternatives to this oxide include conducting polymers, carbon nano-tubes and graphene but their flexibility is offset by low conductivity. Metal nanowire-based electrodes can offer sheet-resistances of less than 10Ω/sq, plus 90% transmission, due to the high conductivity of the metals. The nano-wires must however be defect-free and possess conductivities which are close to their bulk values. They must also be as long as possible in order to minimize the number of wire-wire junctions, and have a low junction-resistance. A simple fabrication process was described which satisfied all of these requirements and produced a type of transparent conducting electrode that offered a sheet-resistance of some 2Ω/sq and 90% transmission, plus remarkable mechanical flexibility during stretching and bending. The electrode comprised a free-standing metallic nano-trough network and was produced by electrospinning and metal-deposition. Electrospinning and electroless deposition were used[10] to create interconnected ultra-long metal nano-wire networks. The process was scalable and involved ambient temperatures and pressures. When using either silver or copper nano-wire, the resistances and transmittances attained some 10Ω/sq and 90%, respectively.

Density functional theory was used[11] to investigate the optical properties of χ^3-borophene monolayers, nano-ribbons and nano-tubes as potential so-called transparent metals. The optimized lattice parameters of pristine χ^3-borophene monolayers were $a = 2.91$Å and $b = 4.45$, with the angle between a and b being 70.92°. This phase was not dynamically stable in the free-standing state, due to phonon instability. It could be trapped in a metastable state, and stabilized by charge-transfer to the underlying substrate. The cohesive energy of 25% oxygen-defected borophene, χ^3-borophene nano-ribbon, zig-zag (12,0) and (16,0) and armchair (6,6) and (8,8) nano-tubes were compared. The adsorption of oxygen on the monolayer was barrier-less, and reduced stability. The material had a very high optical transparency, and zig-zag nano-tubes exhibited 100% optical transmission in visible, ultra-violet and some infra-red ranges. There was a marked decrease in the absorption coefficient of zig-zag nano-tubes, as compared to that of monolayers. By increasing the nano-tube diameter, and changing its chirality, the transmission spectrum could be tuned. Zig-zag nano-tubes with larger diameters exhibited essentially 100% optical transmission in the infra-red, visible and ultra-violet ranges. The combination of high optical transparency and the material's metallic nature promised application in photovoltaic devices.

The photoelectric potential of a borophene|TiO_2 bilayer was estimated[12] by analyzing the atomic-level interactions of the interface and by using first-principles density functional theory. Rutile (001) was combined with χ^3 borophene in a nano-composite with high interfacial coupling. Small interplanar distances and high adhesion-energies were found for optimized structures. The interplanar distances and work function were 1.38Å and 4.49eV. The structure of the interface between χ^3 borophene sheet and TiO_2 substrate changed upon contact with a binding energy of 1.67eV, and was stable. A band-gap of about 0.369eV, the electronic band structure and the density-of-states indicated a potentially superconducting nature. It was possible to modify the work-function of both materials. Such a system was able to absorb visible light, given that it had the earliest absorption peaks in the visible and ultra-violet ranges. This allowed partial transparency, and appreciable photoconductivity occurred at photon energies of less than 12eV. The opto-electronic properties of the borophene|rutile materials could be tuned.

A method for improving the flexibility of transparent electrodes was based[13] upon reducing the crack length, under cyclic folding fatigue conditions, in a metal-polymer hybrid nano-structure comprising indium tin oxide, silver and a fluoropolymer. Solution-processed transparent electrodes were produced[14] which consisted of random meshes of metal nano-wires that offered a transparency which was at least equivalent to that of metal-oxide thin films having the same sheet resistance. Organic solar cells which were deposited on these electrodes performed as well as devices which were based upon a conventional metal-oxide transparent electrode. In order to guarantee an adequate bending fatigue lifetime, precise evaluation of the bending strain and the change in electrical resistance is required[15]. A previous investigation of the bending strains of copper thin films on flexible polyimide substrates of various thicknesses involved using monolayer and bilayer bending models and monitoring of the electrical resistance of the metal electrode during bending fatigue tests. In the case of a thin metal electrode, the bending strain and fatigue lifetime were similar, regardless of the substrate thickness. In the case of a thick metal film, the fatigue lifetime was affected by the bending strain in the metal electrode, depending upon the substrate thickness. In order to determine the exact bending-strain distribution, finite-element simulations were used and the bending strains in thin and thick metal structures were compared. For thick metal electrodes, the real bending strain which was deduced using a bilayer model or simulation yielded values which were very different to those found using a simple monolayer model. Indium tin oxide has a high transmittance in the visible range and a very low resistance at a thickness of about 200nm. This material thus predominates in opto-electronic devices which are based upon rigid substrates. On the other hand it involves a high sputtering cost, suffers from poor flexibility and fragility while, moreover, indium is both rare and toxic. The metal-polymer hybrid nano-structure was varied by sputtering the silver using powers ranging from 20 to 50W. The changes in resistance following 100000 folding cycles at a peak strain of 2.5%, ranged from 4.57 to 17.9%. The changes in the case of oxide|silver|oxide thin-film electrodes without the metal-polymer hybrid nano-structure, increased markedly to 11228%. By introducing the metal-polymer hybrid nano-structure, the crack-length was reduced to 8.76 to 88.1µm. In the case of the oxide|silver|oxide thin-film electrodes it was greater than 945µm. Some 4N-purity silver and 4N-purity $In_2O_3:Sn_2O_3$ sputtering targets were used to form a nano-structured conducting layer. In order to vary the number-density of agglomerated silver nano-particles, the direct-current sputtering power was varied from 10 to 50W. The upper and lower indium tin oxide layers were deposited using a direct-current pulse power of 100W to produce thicknesses of 31.5 and 13.5nm, respectively. The intermediate 6nm silver layer was deposited using a direct-current power of 35W. The control sample (without the metal-polymer hybrid nano-structure) prepared in the same way. All of the samples were annealed (250C, 0.5h, air). The sheet resistances of the flexible electrodes were measured using 4-point probe techniques. A hybrid structure was observed, in which the fluoropolymer co-existed in the silver agglomerate layer due to penetration of silver nano-particles into the fluoropolymer layer. The oxide|silver|oxide thin-film electrode exhibited the optical transmittance which was typical of such transparent electrodes, in that transmission decreased with increasing wavelength in the visible region. This was attributed to an increase in the reflection or absorption of light due to surface plasmon resonance. The transmission of the metal-polymer hybrid electrodes was lower than that of the oxide|silver|oxide thin-film electrode as the nano-agglomerated silver, which caused surface plasmon resonance to occur, was further inserted. As the sputtering power was increased from 10 to 50W, the transmission at 550nm decreased from 74% to 66% as the surface coverage of the nano-agglomerated silver array increased. The resistance values of the metal-polymer hybrid electrodes were lower than those of the oxide|silver|oxide thin-film electrodes, due again to the insertion of silver|fluoropolymer nano-structured arrays, which interfered with electron flow. The resistances of metal-polymer hybrid electrodes with higher sputtering powers increased because of an increase in the surface coverage of the nano-structured array. It remained below 1000Ω/sq; at least comparable to those of previously reported flexible indium tin oxide electrodes having nano-structures of the order of 100 to 10000Ω/sq. The metal-polymer hybrid electrodes which corresponded to sputtering powers of 10 to 35W (table 1) had resistances of the order of 400 to 700Ω/sq. The 100000 cyclic-folding tests were performed using a fixed bending radius of 1mm; corresponding to a compressive peak-strain of 2.5%. The oxide|silver|oxide thin-film electrodes did not exhibit good flexibility under severe cyclic folding conditions. Following 100000 cycles, the metal-polymer hybrid electrodes exhibited better flexibility. The metal-polymer hybrid electrodes were much more

Materials Research Forum LLC
https://doi.org/10.21741/9781644903476

resistant to folding-cycle fatigue, with the 20-, 35- and 50W-prepared samples having the most stable flexibilities and a resistance change of 6.49 to 15.1%. The results were explained by comparing the surface morphologies. The figure-of-merit was taken to be εN which, in the present study, gave 2.5% x 10^5 = 2500. This represented the degree of cyclic-folding fatigue. Most previous studies of flexible electrodes had yielded figures of less than 1000, and sometimes as low as 50. Of the metal polymer hybrid electrodes, the 20W-prepared sample had the best flexibility, and a change in resistance of 4.57% under cyclic folding fatigue. The metal polymer hybrid electrodes also had shorter crack-lengths and greater radii-of-curvature at the crack-tip than did the thin-film electrode. This was because of a ductile nano-structure which increased the toughness. Crack propagation was thereby suppressed due to low stresses at the crack tip … resulting in an improved flexibility. The crack length could be adjusted by varying the density of the metal polymer hybrid nano-structure. The latter played a pivotal role in improving the flexibility. Most of the oxide films in oxide|metal|oxide films are deposited by using magnetron-sputtering or thermal evaporation methods, where high-power ion-bombardment or high temperatures would degrade device performance. In order to create a semi-transparent solar cell, an oxide|metal|oxide has to be deposited on top of the semiconductor layer. It is essential to reduce damage to the semiconductor, in order to reduce the possible trapping of photogenerated charges. Magnesium and gallium co-doped ZnO has a wider spectral transmittance and causes less damage to the underlying functional layers in opto-electronics, but its conductivity is limited. In order to ensure a low sheet-resistance, the layer thickness has to be several hundred nanometres. High-quality magnesium and gallium co-doped films were deposited[16] using reactive plasma deposition at room temperature, without intentional heating, and provided broadband transmission and ultra-thin pure silver films were prepared by magnetron sputtering at room temperature. Compared with the single MGZO, MGZO|Ag|MGZO multilayers effectively improved thin-film conductivity while maintaining high transmittances. The transfer matrix method was used to determine the optimum thickness of each layer in the oxide|metal|oxide, and there was excellent agreement between the simulation and experimental results. An MGZO|Ag|MGZO (40|9.5|45nm) transparent electrode on a glass substrate presents an average transmittance of 87% (including glass) over a spectral range of 400 of 800nm (relative transmittance, 94.7%) and a sheet-resistance of 10Ω/sq.

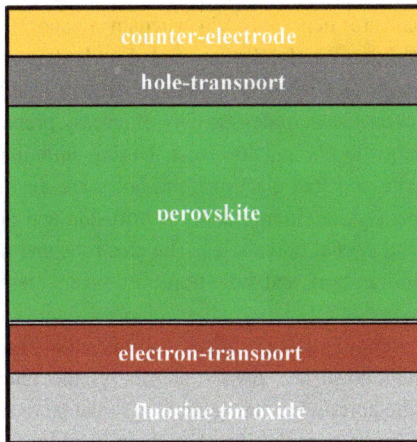

Figure 2. Generic layout of a perovskite solar cell

A semi-transparent perovskite solar cell with an oxide|metal|oxide top electrode exhibits a photoelectric conversion efficiency of about 11%. This work provided an insight into the growth of high-quality oxide|metal|oxide films combining the RPD-TCO technique of high-rate deposition and low ion bombardment with ultra-thin silver films at room temperature, which thus encouraged oxide|metal|oxide use in more diverse opto-electronic devices. A high threshold-thickness of the metal layer in oxide|metal|oxide thin films leads to strong reflectance, especially in the near-infrared region; thus limiting broad-spectrum applications.

A typical perovskite solar cell consists of a transparent substrate, a perovskite layer situated between electron-transport and hole-transport layers, and a metallic upper counter-electrode (figure 2). The transparent substrate is usually made from fluorine-doped or indium-doped tin oxide. In mesoscopic and n–i–p type perovskite solar cells, illumination first impinges upon the electron-transport layer while, in inverted planar heterojunction structures, illumination first impinges upon the hole-transport layer. In mesoscopic and planar heterojunction cells therefore, the interfaces of the cells are mainly of substrate|electron-transport-layer, electron-transport-layer|perovskite, perovskite|hole-transport-layer and hole-transport-layer|counter-electrode type. In the case of inverted planar heterojunction solar cells, they are of substrate|hole-transport-layer, hole-transport-layer|perovskite, perovskite|electron-transport-layer and electron-transport layer|counter-electrode type. Most perovskite solar cells have gold or silver counter-electrodes which are deposited via vacuum evaporation or electron-beam sputtering, although aluminium, molybdenum, nickel and silver-aluminium alloys have also been used for that purpose. Carbon-based materials have also been considered to be very suitable for use as counter-electrodes, as they offer easy preparation, good chemical stability, compatible energy levels and low cost. During illumination, the photons are absorbed by the perovskite and free electrons and holes are excited. Free carriers then first travel through the perovskite via migration or diffusion and may then recombine or be scattered by defects and crystal boundaries. The electrons and holes are subsequently gathered by the electron-transport and hole-transport layers, respectively. In order to ensure a high power-conversion efficiency, all of the photo-induced carriers should reach the external load, but various defects arising from lattice mismatch, energy-level misalignment and thermal noise are always present in the interfaces. The electron-transport layer prevents the holes which are generated in the perovskite from reaching the cathode. The latter can also be considered to be a hole-blocking layer. When the lowest unoccupied molecular orbital of the electron-transport layer is far lower than that of the absorber, there is a decrease in the open-circuit voltage. The highest occupied molecular orbital of the hole-transport layer should be just slightly higher than that of the perovskite, in order for it to be possible to collect holes and block electrons. The electron-transport-layer|perovskite interface thus plays an essential role in determining the open-circuit voltage. Meanwhile the perovskite|hole-transport-layer interface is more important to the control of the photo-current. The work functions of adjacent layers in perovskite solar cells, which should be closely matched, can however be adjusted by means of interface engineering. In early work, organic lead halide perovskite solar cells were prepared which included a transparent sputtered indium tin oxide upper electrode[17]. A power-conversion efficiency of 1.5% was measured for transparent cells with a thin molybdenum oxide buffer layer and indium tin oxide electrode. An efficiency of 2% was found even in cells with the tin oxide electrode sputtered directly onto the organic charge-transport layer.

Table 1. Effect of cyclic-folding fatigue deformation on thin-film and metal-polymer hybrid electrodes

Sample	R(Ω/sq)	Transmission(%)	Crack-Length(μm)	$\Delta R/R_0$(%)
thin-film	48.5	78.0	>945	11228
10W MPH	476	74.0	88.1	166
20W MPH	565	72.6	54.8	4.57
35W MPH	676	67.5	18.3	13.9
50W MPH	983	66.0	8.76	17.9

A zinc-doping method was proposed[18], when using single-target sputtering technology, in order to carry out the growth of zinc-doped silver thin films and to introduce trace amounts of oxygen in order to obtain ultra-thin silver|Zn(O) films having a thickness of less than 5nm. This markedly improved the broad-spectrum characteristics of oxide|metal|oxide films. The use of heterogeneous metal and gas-doping technology promote the occurrence of 2-dimensional continuous film growth. By combining an ultra-thin silver|Zn(O) layer with a magnesium and gallium co-doped ZnO film grown by reactive plasma deposition, a broad-spectrum compound thin film could be prepared which offered an average transmittance of 91.6% at wavelengths ranging from 400 to 1200nm; together with a low sheet-resistance. Broad-spectrum organic solar cells which were based upon these composite electrodes offered a power-conversion efficiency of 15.35%; better than that of devices based upon single-layer oxide electrodes. Aluminium-doped-ZnO|Ag|Al-ZnO multi-layer coatings, 50 to 70nm thick, were grown[19] at room temperature onto glass substrates to a silver-layer thickness of 3 to 19nm, by using radio frequency magnetron sputtering. The thermal stability of the compositional, optical and electrical properties of the structures were investigated up to 400C as a function of the silver film thickness. An Al-ZnO film as thin as 20nm was an excellent barrier to silver diffusion. The inclusion of a 9.5nm silver layer in the transparent conductive oxide led to a maximum enhancement of the electro-optical characteristics.

Transparent conductors are naturally characterized by both a high light-transmission and very high direct-current conductivity[20]. Many applications also require that they be mechanically strong and flexible. They are broadly of two types: uniform and non-uniform. The former include conventional metals, or electron plasmas, with the plasma frequency located within the infrared frequency range, such as transparent conducting oxides. There are also ultra-thin metals with a wide plasma frequency. The physics of non-uniform transparent conductors is more complicated and may involve transmission-enhancement due to refraction (including plasmonic) and exotic electron transport effects such as percolation and fractal-like. There are non-uniform metallic films such as random metallic networks, and the transparency of such networks might be increased beyond the classical shading limit by plasmonic refractive effects. The conduction depends strongly upon the network type, including individual metal nano-wires where the conductivity depends upon the inter-wire contact, and percolation effects. Uniform metallic film networks are free from percolation effects and contact problems. In applications which require high-quality electric contact of the transparent to an active substrate such as a solar cell, the network behaviour can be optimized by imposing a quasi-fractal network structure. In periodic metallic networks, active plasmonic refraction can lead to so-called extraordinary optical transmission. A study was made[21] of the mode-characteristics of multi-layer metal-dielectric nano-film structures which could be described as being coupled-plasmon resonant waveguides; and in turn a special case of coupled-resonator optical waveguides. Like a photonic crystal, the metal-dielectric was periodic, but fields were evanescent everywhere in the latter structure, as in a nano-plasmonic structure. The transmission coefficient still underwent a periodic oscillation with increasing number of periods. Due to surface-plasmon enhanced resonant tunnelling, 100% transmission periodically occurred at certain thicknesses of the structure, depending upon the wavelength, the lattice constants and the excitation conditions. A transparent material could thus be composed of non-transparent materials by alternately stacking thin layers of various materials. So-called smart transparent conductors may incorporate added properties such as light interference, metamaterial effects and in-built semiconduction. An understanding of ultra-thin metal-film opto-electrical properties as a function of thickness is necessary for the design of transparent conductors. Combining the film with dielectric coatings can also improve not only optical transmission but also mechanical flexibility. There is a critical thickness below which the film changes from homogeneous to inhomogeneous[22]. In the homogeneous regime the metal film becomes continuous, and free from voids. In the inhomogeneous regime the film can be treated as a metal-void composite material. In the ultra-thin film case, the two regimes become less distinct. Below the critical thickness, the opto-electrical properties markedly deteriorate. It is therefore necessary to choose a film thickness which is above this critical value in order to achieve a figure-of-merit for the transparent conductor. Resistivity modelling shows that grain-boundary scattering is the predominant mechanism in the deterioration of the opto-electrical properties of intrinsic metal films. Anti-reflective coatings further improve the optical transmission of metal-film stacks (table 2). The design of such stacks is not limited to the transparency at visible wavelengths, but also extends into the near-infrared range. Because metallic thin film is malleable, crack-propagation is more difficult than it is in indium tin oxide conductors. The presence of a thin metallic layer between indium tin oxide layers can partially absorb bending deformation-energy via shear deformation of the metallic layer. A hierarchical multilayer structure can change the direction of crack propagation and delay electrode degradation. Flexible transparent electrodes were created[23] which featured a metal mesh that was fully embedded in a flexible substrate. The embedded nature of the metal-mesh electrodes provided surface smoothness, mechanical stability under high bending-stresses and strong adhesion to the substrate. The fabrication process involved electrodeposition and permitted the preparation of high aspect-ratio (thickness to linewidth) metal mesh. This greatly improved the conductivity without losing much transparency. Prototype flexible transparent electrodes had transmittances higher than 90% and sheet-resistances of less than 1Ω/sq, together with figures-of-merit of up to 1.5×10^4.

Table 2. Electrical resistance and optical transmittance of indium tin oxide and possible flexible transparent conductors on a flexible substrate

Material	Resistance(Ω/sq)	Optical Transmittance(%)
indium tin oxide	13.4	85.7
metal nano-wire	12.5	85.6
dielectric/metal/dielectric film	18.6	88.4
metal mesh	8.2	88
carbon-nanotube/Ag-nanowire	50	94
graphene/metal composite	86.8	>90

It is a factoid known to every schoolboy that gold can be beaten-out to such an extent that visible-light sources can be seen through it. The usual means for seeing through metals is the use of X-rays. It is perhaps worth mentioning here in passing that the direct real-time microscopic observation of changes within metals, such as solidification, is possible by means of synchrotron X-ray imaging. Although X-ray lenses exist, and can be inserted into the beam-path, the basic microscopic technique is actually projection. By using a scintillator screen to convert the X-rays into visible radiation, a projected image can be examined in detail by using an optical microscope. Because the X-rays can penetrate quite thick samples, it is possible to exploit some of the techniques of conventional medical radiography. That is, a large number of 2-dimensional projected images of a sample, seen in 500 to 2000 orientations, can be combined so as to produce a high-resolution 3-dimensioal tomogram. It is also possible to exploit some of the tricks which are used by optical microscopists. The X-ray image-contrast weakens with increasing photon-energy, but high-energy photons are of course the very ones which are required in order to penetrate a metallic sample. The familiar optical Zernike phase-contrast technique can enhance the contrast by converting phase modulations in the image into detectable amplitude modulations[24,25,26,27,28,29]. The focus here however is more on the schoolboy case, and the use of films which are so thin as to be effectively transparent to the naked eye.

The spectroscopic ellipsometry method for the characterization of thin metal films and the optical design procedure for optimizing the transmittance of thin metal film conductors were reviewed[30] with regard to their use in solar cells, organic light-emitting diodes and transparent electromagnetic interference coatings. The electron transport of ultra-thin metal film exhibits a transition at a critical thickness. Due to the rapid increase in electrical resistivity below the critical thickness, film of less than this thickness becomes less suitable for use as a transparent conductor. Note that this critical thickness is distinct from the percolation threshold. Above the critical thickness, where the ultra-thin metal film is still homogeneous, the optical transmission of the film increases with decreasing thickness. This continues up to the critical thickness, below which there is no further gain in optical transmission in spite of the reduction in film thickness. When below the critical thickness, plasmonic resonance absorption increases due to an island-like morphology of the inhomogeneous metal-insulator medium. This implies that the Haake figure-of-merit attains a maximum near to the critical thickness. It is thus desirable that the metal film should be as thin as possible, in order to maximize optical transmission, but not so thin that electrical and optical properties begin to deteriorate. Due to the very reflective nature of thin metal films at optical frequencies, light transmission through the film is still limited. Even when the thickness is extremely reduced, light reflection at the metal-air/metal-substrate interface can be significant; especially at longer wavelengths. One method of increasing transmittance is to use an anti-reflective dielectric coating on each side of the metal film, leading to 3 reflection coefficients. That is, r_1 is the reflection-coefficient of the incident light which is reflected at the air/dielectric interface, r_2 is the reflection coefficient when light passes through the dielectric layer and is reflected back at the dielectric/metal interface and r_3 is the reflection coefficient when light passes through the metal layer and is reflected back. This

Transparent Metals   Materials Research Forum LLC
Materials Research Foundations **174** (2025)   https://doi.org/10.21741/9781644903476

includes reflection from the metal/dielectric and dielectric/substrate interfaces. Dielectric|metal|dielectric structures should be optimized so that the sum of r_1, r_2 and r_3 equals zero. This ensures minimum reflection and maximum transmission. Destructive interference occurs when $r_1 + r_3 = r_2$ and completely suppresses overall reflection. Within the visible range, r_1 and r_3 are usually in phase, but are both out-of-phase with r_2, leading to essentially complete destructive interference. In the near-infrared range, because of the high extinction for propagation through the metal layer, the amplitude of r_3 markedly decreases due to absorption. A sharp phase-shift at the metal|dielectric interface causes r_3 no longer to be in phase with r_1, such that $r_1 + r_3 \neq r_2$. This means that a dielectric|metal|dielectric structure has a limited number of interfaces and propagation media in which both can modulate the reflection coefficients. The addition of layers below or above the metallic layer can introduce extra r_1, r_2 and r_3 components which suppress reflection by destructive interference in the near-infrared range. Because of extra components in each reflection coefficient at the additional interfaces the amplitudes of r_1 and r_3 are increased and the net phase-shift between r_1 and r_3 is no longer entirely out-of-phase. The destructive interference of $r_1 + r_3 = r_2$ remains satisfied and leads to an anti-reflection effect which extends into the near-infrared region. In order to suppress further reflection and maximize transmission, more dielectric layers might well be added. In flexible transparent conductors, electrical failure can result from large mechanical deformations and crack propagation across the conductor during repeated bending or straining. Dielectric|metal|dielectric multilayer structures not only improve light transmission but also exhibit great endurance in bending tests. The good flexibility of dielectric|metal|dielectric based transparent electrode can be attributed mainly to the ductile metal interlayer. This ensures electrical conductivity even when the dielectric is deformed far beyond its fracture strain. Flexible electrodes which are based upon metal thin films on elastomeric substrates can suffer from a complete unexpected electrical disconnection because of mechanical fracture of the metal. It was shown that the strain-resilient electrical behaviour of thin-film metal electrodes under multi-mode deformation can be improved by using a 2-dimensional interlayer[31]. The insertion of an atomically thin interlayer leads to continuous in-plane crack deflection in thin-film metal electrodes and leads to a so-called electrical ductility in which the electrical resistance gradually increases with strain so as to create extended regions of stable resistance. Two-dimensional interlayer electrodes can maintain low electrical resistance at strains below which conventional metal electrodes would disconnect. Indium tin oxide is the most important single-layer transparent conductor and can offer a resistivity below $20\Omega/sq$, but it is necessary to maintain a low substrate temperature during its deposition onto polymer films. One solution is to replace indium tin oxide by a 3-layer stack consisting of a thin silver layer plus two embedding transparent layers[32]. The key concept was to introduce two interlayers into the calculation model. The interlayers represented the properties of the boundary of the silver layer, and calculations which were based upon this model agreed very well with measurements. Optical absorption spectra indicated qualitative agreement with the behaviour of surface plasmons. The roughness of the layers and the dielectric constant of the embedding materials was represented by the model. Two types of failure occur during bending: fracture and fatigue. Bending-mode tests are commonly used to evaluate the strain tolerance of flexible electronics. Copper-based materials are attractive due to their electrical, mechanical and thermal properties, and copper nano-wires are typically prepared hydrothermally. But unlike silver nano-wires, copper nano-wires are very vulnerable to degradation by oxidation. As well as metal-based nano-wires, metal grids are good candidate materials because of their high electrical conductivity, optical transparency and mechanical resistance. Nano-scale metal grids also permit the exploitation of polarization selection and plasmonic excitation. The fatigue behaviours of silver nano-wires, metal grids and $Al_2O_3|CuAg|Al_2O_3$ structures are much worse than that of dielectric|metal|dielectric structures. In silver nano-wires, the change in resistance during bending strain is closely related to the failure of the individual nano-wires. These nano-wires exhibit the characteristics of a size-dependent ductile-brittle transition and brittle fracture can occur at small deformations. Transparent electrodes comprising single-component less-than-10nm metal films were prepared[33] by sputtering-deposition. The optical transparency of the chromium and nickel films was comparable to that of indium tin oxide in the visible and near-infrared range of 0.4 to 2.5μm, but could be much higher in the ultraviolet range of 175 to 400nm and the mid-infrared range of 2.5 to 25μm. The deposited films were uniform and continuous over the 10cm substrate and this was confirmed by the low electrical resistivity. A substrate-embedded thick (350nm) and thin (30nm) 3-dimensional metal grid mesh structure with a large area was prepared by secondary sputtering[34]. This offered a sheet-resistance of 9.8Ω/sq and a transmittance of 85.2%, plus a high stretchability with no great change in resistance for applied strains of less than 15%. The mesh had a sub-micrometre period, a root-mean-square roughness of about 5nm and exhibited strong adhesion to polymer substrates. These properties were attributed to the substrate-embedded 3-dimensional structure of the electrode, which could have a high aspect-ratio and high resolution over large areas. A scalable solution-based method was used[35] to produce silver micro/nano-wire networks for transparent conductors. By exploiting self-cracking, and using water-soluble acrylate co-polymer film as a photo-resist mask, the use of photo-mask fabrication, vacuum and lithography was avoided. An increase in adhesion and a decrease in roughness of the metal networks was demonstrated by depositing metal into the regions created by the glass-etching step.

Transparent Metals   Materials Research Forum LLC
Materials Research Foundations **174** (2025)   https://doi.org/10.21741/9781644903476

The networks had record-making figures-of-merit. Metal grids are prone to fatigue damage because stress concentrates in the narrow conductive path during cycling, and makes it difficult to resist fracture and necking. Dielectric|metal|dielectric with an Al_2O_3|CuAg|Al_2O_3 multilayer suffers failure earlier than does dielectric|metal|dielectric with an ITO|CuAg|ITO structure. This can be attributed to the low toughness of Al_2O_3. The key parameter which characterizes brittle thin films on polymeric substrates is the fracture toughness. The elastic modulus of Al_2O_3 is 200GPa, and is thus higher than that of indium tin oxide (150GPa), but the former has a toughness of only 0.24 to 1.20MPa√m while that of indium tin oxide is 2.50 to 2.70MPa√m. An ITO|CuAg|ITO structure can thus absorb fracture-energy and deform plastically without fracture. This suggests that the ductility of the CuAg layer is crucial for fatigue-resilient electrical performance during cyclic bending. An electrically conductive path exists even when the indium tin oxide is strained beyond its failure limit. When a crack tip approaches the CuAg layer, the crack is suddenly deflected into the lateral direction and continues to propagate along the metal|ITO interface. This creates an unusual step-like crack, and the deviation in crack direction can be attributed to competition between the direction of maximum mechanical driving force and the weakest structural pathway. The ductile layer of a thin metal film presumably improves the mechanical properties of the entire structure, due to internal interfaces. The CuAg layer plays an important role in preventing rapid crack propagation because it absorbs energy via large shear deformations, and strain-energy transfer to plastic deformation. This leads to the characteristic step-like fracture. With regard to electrical conductance, vertically penetrating cracks are the worst because they severely damage the conducting material. In the presence of a thin metal layer, the crack is diverted and overall conduction is maintained. The toughening mechanism of the dielectric|metal|dielectric structure is associated mainly with the 2-dimensional deflection of cracks which bypass the thin CuAg layer. The latter dissipates strain energy arising from the occurrence of plastic deformation and greatly decelerates crack growth. Within the metallic layer, bond-rupture plays a crucial role in resisting crack growth because fracture energy is closely associated with plastic deformation near to the crack tip. When pure indium tin oxide is subjected to bending, the cross-section of the entire stack remains planar and has a single neutral plane after deformation. Multiple neutral planes appear upon inserting a ductile metallic layer between the two indium tin oxide layers, and the oxide layers bend individually during bending. They are not rigidly coupled because the transverse sectional plane of the metallic layer deforms so as to accommodate the sheer stress. The multiple neutral planes permit the transfer of strain energy to the middle layer so that the maximum bending strain can be reduced with increasing compliance of the structure. The reduced strain also decreases the probability of crack initiation.

Depending upon the substrate surface-condition and the type of metal, the deposition of thin films can follow various paths. In the Volmer-Weber case, island growth occurs when the interaction between neighbouring metal atoms is stronger than the interaction between the substrate and metal atoms[36]. In the Frank-Van der Merwe case, layer growth occurs when the interaction between the substrate and the metal atoms is stronger than the interaction between the neighbouring metal atoms. This is an entirely 2-dimensional growth mode in that the metal atoms bond to atoms of the substrate, and a first layer is completely finished before the next layer starts to grow. In the Stranski-Krastanov case there is mixed, island and laminar, growth in which 1 or 2 monolayers are deposited before individual islands appear on them. The growth of ultra-thin metal films tends to follow the Volmer-Weber path, with individual 10 to 20nm metal islands later converging to form a continuous film. Ultra-thin metal film are added to transparent electrodes in a step before indium tin oxide (table 3), and are a viable replacement for the oxide. Ultra-thin metal films offer high transmittance but Volmer-Weber growth produces non-continuous rough films having poor electrical conductivity. An associated microstructure on the surface increases light scattering at the interface and impedes transmittance of transverse currents, thus being detrimental to the optical and electrical properties of the thin metal film. Deposition of a suitable thin metal film can be facilitated by increasing the surface free energy of the substrate or increasing the binding energy between the metal and the substrate. This creates smooth ultra-thin continuous films. It was demonstrated[37] that ultra-thin ultra-smooth low-loss silver films could be created by using a very thin germanium layer as a wetting material and applying a rapid post-annealing treatment. The addition of the germanium wetting layer greatly reduced the surface roughness of silver films which were deposited onto a glass substrate by means of electron-beam evaporation. The percolation threshold of the silver films and the minimum thickness of a uniformly continuous silver film were markedly reduced by using the germanium wetting layer. Rapid post-annealing reduced the loss of ultra-thin silver film to the optimum value which was allowed by the quantum size effect in smaller grains. Transparent conductive electrodes were constructed[38] which consisted of patterned few-nanometre silver films on zinc oxide coated rigid or flexible substrates. The grid lines were entirely continuous and 8.4nm thick. Due to the high transparency of the grid-lines and to the spacing, electrodes with an opening ratio of just 36% offered an average optical transmittance of up to 90% in the visible regime. This broke the optical limits of the unpatterned and thick-grid equivalents, where the transmittance was governed by the opening ratio. The small value of the latter led to a sheet-resistance of

Materials Research Forum LLC

https://doi.org/10.21741/9781644903476

21.5Ω/sq. The figure-of-merit of up to 17 was higher than that of the non-patterned-film equivalent. The ultra-thin electrode, firmly attached to the substrate, was mechanically more flexible than the film equivalent or indium tin oxide.

Table 3. Properties of transparent metal electrodes evaporated onto various seed layers

Seed-Layer	Film	Transmittance(%)	λ(nm)	Roughness(nm)	R(Ω/sq)
Cr	Au	70	350-1200	-	16.3
Cr	Au	60	-	27.9	
Ge	Ag	85.33	550	0.11-0.16	25
Au	Ag	85	550	-	16
Ag	Au	80	-	16	
Ni	Ag	75	3.9	11	
Al	Ag	87	550	-	19.5
Cu	Ag	62.77	400-1200	0.204	19
Cu	Au	75	550	5.4	16
PEI	Ag	80	550	0.23	9
PAI	Ag	87.4	550	0.768	15.1
PEI	Ag	69.7	550	<1	6.3
PVK	Ag	>85%	-	<10	-
SU-8	Au	72	550	0.35	23.75
SU-8	Au	80	550	0.575	19
MUA	Ag	78	400	0.95	13.59
PEI	Ag	80	550	0.15	9
PAA	Ag	75	550	-	10
PVP	Ag	70	550	-	19
PMMA	Ag	45	550	-	
PFN	Ag	54.3	1.3	9.4	-
Alucone	Au	77.8	0.18	30	-
PMMA/TMA	Au	84.25	550	0.566	18.19
MPTMS	Ag	76	0.5	6	-

A general method was developed[39] for creating highly transparent ultra-thin silver films on glass or plastic substrates by using a self-assembled monolayer modified ZnO|Ag|ZnO tri-layer structure. The resulting films offered very low surface roughness, a transparency of greater than 80% between 400 and 600nm and a surface resistance of 8.61Ω/sq. They also had superior mechanical properties.

A low-voltage transparent metal-semiconductor-metal ultraviolet photo-detector was demonstrated[40] which was based upon asymmetrical interdigitated gold electrodes with a thickness of much less than 10nm. An in-plane metal-semiconductor-metal configuration is favourable for higher transparency when transparent conductive electrodes are on the same surface of the semiconductor layer. Transparent conductive oxide | metal | transparent conductive oxide multilayer structures provide optical and electrical characteristics which are better than those of possible by using a single-layer oxide or metal electrode[41]. They can also be deposited at low temperatures onto plastic substrates. Organic electronics which require low deposition-temperatures have the best chance of being transferred from glass to plastic substrates. Interdigitated fingers between the electrodes greatly improve the transportation and collection of photocarriers. In this case the electrodes must be highly transparent and indium tin oxide or ultra-thin silver films are therefore normally used. The preparation method was simple and scalable. Pieces of silica glass were used as transparent substrates for the deposition of gold to thicknesses of up to 20nm, and usually less than 10nm. The photo-detector consisted of a circa 100nm-thick ZnO active layer and ultra-thin gold asymmetrical interdigital electrodes. The ZnO was also a seed layer for the ultra-thin gold-film growth. A 4nm-thick film which was deposited onto a bare silica substrate was not continuous, and included isolated gold nano-particles. When the thickness was increased to 7nm, the isolated nano-particles grew larger and coalesced. The film became continuous at thicknesses of 10 and 20nm. Films which were deposited onto ZnO-coated silica were smoother than those deposited onto bare silica, even at a thickness of 4nm. The gold film became continuous at 7nm, and this was attributed to the good wettability of ZnO. Smoother surfaces made the ZnO-seeded gold films look more transparent. Gold films on bare silica substrates were also more easily damaged, due to their poor adhesion to the substrate. A 7nm-thick film having a visible-range transmittance of 80.4% and a sheet-resistance of 11.55Ω/sq was patterned onto asymmetrical interdigitated electrodes on a ZnO active layer with an average visible-range transmittance of up to 74.3%. The ZnO also facilitated ultra-thin gold-film deposition. A photo-detector with a finger-width ratio of 1:4 offered the best performance. Very low dark currents were found at 0, 0.5 and 1V, and it was very fast under biases of 0.5 and 1V. Gold asymmetrical interdigital electrodes are quite thick, and block the transmission of visible light. The ultra-thin gold interdigitated electrodes rendered these photo-detectors very transparent.

Transparent Metals   Materials Research Forum LLC
Materials Research Foundations **174** (2025)   https://doi.org/10.21741/9781644903476

Highly transparent conductive dielectric|silver|dielectric thin-film multilayers are used as the top electrode of small-molecule organic solar cells[42]. It was shown that 1nm seed layers of calcium, aluminium or gold can markedly affect the morphology of a subsequently deposited silver electrode layer. Wetting by silver of the substrate was greatly improved by increasing the surface energy of the seed material and resulted in improved optical and electrical properties. Thermally evaporated silver on a dielectric typically formed rough and granular layers which were not closed and which were not conductive below a thickness of 10nm. When gold was the seed layer, the silver electrode was a continuous smooth conductive layer down to a thickness of 3nm. When the silver thickness was 7nm, the sheet resistance was 19Ω/sq and there was a peak transmittance of 83% at 580nm. These properties were better than those of silver electrodes without a seed layer and even better than those of indium tin oxide. Top-illuminated solar cells which comprised gold/silver double-layer electrodes offered a power conversion efficiency of 4.7%; close to the 4.6% found for bottom-illuminated devices which used indium tin oxide. Nano-structured transparent metal electrodes have been created by using a displacement-diffusion-etch method[43]. This process could form less-than-20nm thick gold electrodes on various types of substrate. Those which were applied to flexible polyethylene terephthalate substrates imparted high environmental stability and good bendability to flexible indium tin oxide electrodes. Flexible organic solar cells which were made using the gold electrodes (figure 3) offered a similar power conversion efficiency, but greatly increased flexibility, as compared with indium tin oxide devices. The displacement-diffusion-etch process could create the thin gold electrodes at less than 150C. This was done by firstly preparing a sacrificial copper layer on a flexible polyethylene terephthalate substrate by solution-based polymer-assisted metal deposition method. Sub-micron defects were observed on the surfaces of 50nm and 100nm samples, and this could lead to a poor uniformity of the gold film following processing. The copper film became dense and uniform when the thickness was increased to 150nm. Further increase of the thickness to 200nm did not greatly affect the surface morphology, but a thicker copper layer required a longer diffusion time for fabrication. The optimum copper layer was about 150nm thick and adhered strongly to the polyethylene terephthalate substrate via a composite-like copper|polymer interlayer which formed during the polymer-assisted metal deposition. The copper was then partially replaced by gold by using a galvanic displacement reaction in which the copper-coated substrate was immersed in an aqueous solution of HAuCl₄ for 30s. The copper was gradually displaced from the upper surface and the thickness of the copper layer was reduced by about 50% during the displacement. Annealing (150C, 600s) then caused the gold on the upper surface to diffuse into the underlying copper and polymer layers. The annealing led to a

Materials Research Forum LLC

https://doi.org/10.21741/9781644903476

greater particle size and to a greater crystallinity of gold and copper. The copper was finally etched away, by dipping the substrate in $FeCl_3$, to leave an ultra-thin layer of multi-grained gold which adhered to a thin polymer layer on top of the substrate.

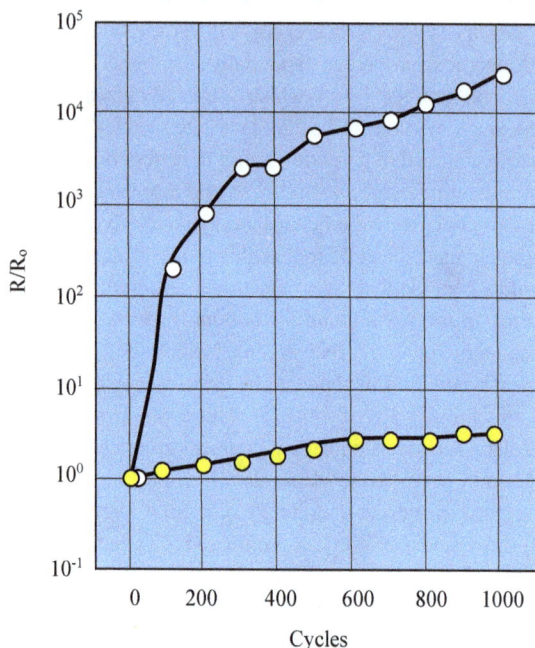

Figure 3. Resistivity change of an ultra-thin gold electrode on a polyethylene terephthalate substrate due to repeated bending to a radius of 1.5mm. Yellow: ultra-thin gold on terephthalate, white: indium tin oxide on terephthalate

Gold electrodes which were prepared using 2.5mM $HAuCl_4$ had the lowest surface roughness, with a root-mean-square roughness of 11.6nm. The surface roughness increased to 12.4, 14.5 and 106.7nm when using 1.25, 5.0 and 10.0mM concentrations, respectively. An increased acid concentration led to an increase in the gold-nanoparticle size and thereby increased the roughness of the films. The reaction was very rapid when the concentration was increased to 10.0mM and the roughness of the film markedly increased due to the presence of large particles and aggregates. Sub-micron defects were present in 5 and 10mM samples, and increased the roughness of the films. The surface

22

roughness of 1.25mM samples was slightly higher than that of the 2.5mM samples because of a poor uniformity of the film. Ultra-thin gold electrodes which were prepared by using 2.5mM acid were the most suitable for use as transparent electrodes for opto-electronic devices. Only about 0.98mg of gold was consumed during the fabrication of 12cm^2 of electrode, and the gold utilization efficiency was about ~47%. The electrical resistance of the electrode increased by less than two times following 1000 bending cycles to a radius of 1.5mm. The resistance of indium tin oxide on the same substrate was anticipated to increase, due to brittleness, by 10000 times following 900 cycles of bending to the same radius. Samples of the present electrodes were subjected to aging for 2 weeks at 120C in a relative humidity of 75%, or to immersion in acid with a pH-value of less than unity. Indium tin oxide or silver nano-wires on polyethylene terephthalate substrates underwent rapid corrosion under the same conditions. There was also a very strong adhesion between the present electrodes and the substrate. The sheet-resistance increased slightly, from 50 to 53Ω/sq. The power conversion efficiencies of solar cells which incorporated the present electrodes retained more than 97% of their original value following 1000 bending cycles to a radius of 1.5mm. Polymer-metal hybrid electrodes[44] offered a bending radius of less than 1mm, a visible-range transmittance of more than 95% and a sheet-resistance of less than 10Ω/sq.[45]

Given that transparency and bendability are desirable features for organic light-emitting diodes, roll-to-roll methods are a promising route for their rapid continuous fabrication. While indium tin oxide is commonly used in organic solar cells as a transparent conductive electrode, the high-temperature processing and poor flexibility of the oxide render it incompatible with large-scale roll-to-roll manufacture of such cells[46]. Instead, MoO$_3$|thin-metal|MoO$_3$ trilayer structures have been used to replace the oxide electrode of the cell. The optical and electrical properties of the trilayer have been found to depend upon the material and the thickness of the intermediate metal layer. A maximum power-conversion efficiency of up to 2.5% was measured under simulated 1-sun AM 1.5 solar irradiation in the case of cells based upon poly(3-hexylthiophene) and [6,6]-phenyl-C61-butyric acid methyl ester, as compared with a maximum efficiency of 3.1% for indium tin oxide devices. The flexibility of the trilayer structure imbued the electrodes with good mechanical flexibility. The efficiency of the flexible device was reduced by only about 6% from its original performance after 500 bending cycles with a radius of 1.3cm. A continuous roll-to-roll method produced[47] transparent conductive flexible plastic which was based upon a metal nano-wire network which was encapsulated between a graphene monolayer and a plastic substrate. The encapsulated structure minimized the resistance of the wire-to-wire junctions and the graphene grain boundaries. This led to a sheet-resistance of 8Ω/sq, with 94% transmittance and great mechanical flexibility. The life of

Materials Research Forum LLC

https://doi.org/10.21741/9781644903476

devices which incorporated it was expected to be up to 10000 cycles. After encapsulating silver nano-wires with a graphene oxide layer[48], the resultant electrodes offered a transmittance of 83.5% at 550nm and a sheet-resistance of 11.9Ω/sq. The reliability, bio-friendliness and long-term stability of the silver nano-wire and graphene oxide hybrid electrodes were also greatly improved. Electrodes which were fabricated on a terephthalate substrate also had an excellent mechanical flexibility. The role played by reduced graphene oxide in chemical stability and its effect upon mechanical reliability was studied[49] with regard to a silver-nanowire|RGO hybrid transparent electrode. Bending fatigue tests of up to 800000 cycles were carried out while monitoring the resistance-change during the fixed uniform bending straining of silver nano-wire networks, with and without a reduced graphene oxide layer. A thin layer of the latter, with an optimized thickness of 0.8nm, when deposited onto the silver nano-wire networks maintained a good reliability of the nano-wire networks, with a fractional resistance increase of 2.7% after 800000 cycles. Use of the layer markedly decreased the oxidation of silver nano-wires, and the bending fatigue properties after exposure to ambient air for 132h at 70C were markedly improved due to suppression of oxide formation on the surface of the silver nano-wires. A very reliable silver-nanowire|RGO hybrid electrode was prepared using mechanical welding and subjecting it to bending-strain so as to form localized junctions without requiring a post-annealing process. Highly transparent co-evaporated Ca:Ag electrodes with a transmittance of 64% at 320 to 800nm (maximum of 70.4% at 380nm) and a sheet-resistance of as little of 21Ω/sq were prepared[50]. Upon increasing the film thickness to 24nm the sheet resistance was further reduced to 12Ω/sq and the transmittance decreased to 58%. The Ca:Ag electrode, with a 2:1 volume ratio, also exhibited good mechanical behaviour during 1000 bending cycles. The electrodes had a root mean square surface roughness of less than 3nm. When encapsulated in ultraviolet-activated getter materials, the coatings exhibited excellent long-term stability over time. Samples with a size of 15cm x 5cm were used for bending tests, with a fixed radius of 5mm and a bending-rate of 30r/min. The bending-angle ranged from 0 up to 90°. Transparent Ca:Ag electrode coatings with a volume ratio of 2:1 or 4:1 were selected for optimization over film thicknesses of 18, 21 and 24nm in order to obtain a transmittance greater 50% and the target sheet-resistance of less than 20Ω/sq. The Ca:Ag coatings with 2:1 ratio had a sheet-resistance of 21Ω/sq at a thickness of 18nm and a sheet-resistance of 12Ω/sq at a thickness of 24nm. The sheet-resistance of 2:1 films decreased in a quasi-linear manner with increasing thickness, but samples with a 4:1 ratio did not follow that trend. The 24nm samples had a higher than expected resistance (70 to 80Ω/sq) and this was attributed to non-ideal deposition. Regardless of the thickness, electrodes with a 4:1 ratio had a higher resistance than did those with a 2:1 ratio (figure

4). There was no appreciable change in the surface roughnesses of the layers (table 4). The surface roughness varied from 1.4 to 3.0nm over the sample area of 5μm x 5μm. In many samples, the peak-to-valley roughness was greater than the film thickness. This suggested that there was a marked dependence of the peak-to-valley roughness upon the local surface quality of the substrate, due to defects or particles at the substrate surface. The 18nm layer with a 2:1 ratio exhibited a maximum transmittance of 70.3% at about 380nm, and an average transmittance of 64% over 320 to 800nm. The transmittance attained 30% for an 18nm coating with a 2:1 ratio, and could be 36% for 21nm film with a 4:1 ratio at 800nm. The results suggested that the Ca:Ag electrodes reacted with atmospheric gases such as H_2O, O_2 and CO_2, and that this reaction markedly affected the optical properties. The transmission and sheet-resistance were therefore monitored for 30 days. The optical transmittance at 550nm greatly increased after a single day of storage in the ambient atmosphere.

Table 4. Surface roughness Ca:Ag thin films on a polyethylene terephthalate substrate

Ca:Ag Ratio(vol%)	Thickness(nm)	RMS(nm)	Peak-Valley(nm)
2:1	18	1.6	30.2
2:1	21	2.2	54.3
2:1	24	1.9	26.2
4:1	18	3.0	28.0
4:1	21	2.8	42.0
4:1	24	1.4	14.2

The transmittance of 18nm film with a 4:1 ratio increased by about 21%, while that of 2:1 ratio film having the same thickness increased by about 27% within 24h. The transmittance of all of the films then remained essentially unchanged, confirming that exposure to the ambient led to rapid reaction between the electrode and the atmosphere over a period of minutes to hours. There were minimal changes in the sheet resistance of Ca:Ag films with a ratio of 2:1 over the test-period of 30 days. On the other hand, the 4:1 films exhibited an increase in sheet-resistance. The resistance of 18nm film increased from 66 to 76Ω/sq, and half of the increase occurred within the first 24h. It was concluded that films with a 2:1 ratio not only had an overall lower resistance of less than

Transparent Metals   Materials Research Forum LLC
Materials Research Foundations **174** (2025)   https://doi.org/10.21741/9781644903476

21Ω/sq, but were also much more stable over time. Bend tests with a fixed bend radius of 5mm and various bending angles were performed on 18nm films. Two different bending directions were used in order to simulate folding and stretching. All of the samples were bent from 0 to 90°, and the resistance was measured for each angle. The 4:1 film alone suffered an increase in resistance, from 263 to 366Ω/sq (40%) at 90° when stretched, suggesting that cracks formed within the thin film. Folding the films caused little change in the resistance. The 2:1 films suffered no appreciable change in properties during opening or folding through up to 90°.

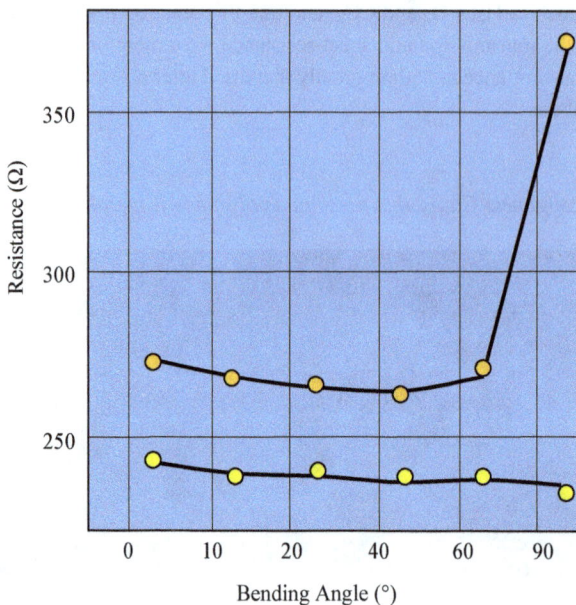

Figure 4. Resistance of 4:1 Ca:Ag coatings on a polyethylene terephthalate substrate bent (5mm) to various angles. Orange: 4:1 ratio (opening), yellow: 4:1 ratio (folding)

During bending to an angle of 60°, Ca:Ag coatings with a 2:1 ratio had a consistent sheet-resistance after 1000 cycles, with a maximum change of 2.8Ω/sq for opening and folding tests. Much higher initial sheet resistances and variations were found for 4:1 films. Encapsulated films which had been UV-treated had almost unchanged optical properties after 4 weeks of exposure to the ambient. Encapsulation without any ultraviolet treatment only temporarily prevented changes in the optical properties. Samples which were exposed to the ambient remained stable for 24h, but longer exposure led to an increase in transmittance like that found for non-encapsulated film.

Table 5. Resistivity and sheet-resistance of thin chromium films

Thickness(nm)	Resistivity(Ωcm)	Sheet-Resistance(kΩ/sq)
5	6.3×10^{-3}	12.6
10	2.25×10^{-3}	2.25
18	2.70×10^{-3}	1.5
22	2.33×10^{-4}	0.106
92	2.11×10^{-4}	0.02293
94	2.16×10^{-4}	0.02298
110	2.83×10^{-4}	0.03093

The development of very thin electron-beam deposited chromium films of high optical transparency was investigated[51]. The electrical and optical properties were determined as a function of the thickness and deposition rate. For a given film thickness the absorption coefficient increased as the deposition-rate was decreased. Films having a thickness of less than 10nm had an average transmittance of more than 60% at 400 to 1500nm. The resistivity varied from 6.3×10^{-3} to $2.83 \times 10^{-4}\Omega$cm for coatings with a thickness of 5 to 100nm. The imaginary part of the permittivity decreased with decreasing film thickness. A smooth surface was observed in every case. The transmittance was lower at shorter wavelengths and increased at higher wavelengths. The transmittance was greater than 65% for films with a thickness of 10nm. Film with a thickness of 5nm had a transmittance of 67 to 80% in the visible range, and a transmittance greater than 80% in the near-infrared range. A decrease in the film thickness led to a fall in the refractive index. The values of the latter ranged from 2.5 to 3.5 for thicknesses of 16 and 35nm, and from 1.47 to 1.64 for a 5nm coating at wavelengths of 400 to 1500nm. The resistivity decreased for thicker films (table 5). A sudden increase in the resistivity occurred for films which were thinner than 22nm. As a comparison, the sheet-resistance of indium tin oxide films with a thickness of 50 to 100nm ranged from 2 to 6kΩ/sq.

The power-conversion efficiency of semi-transparent plastic solar cells could still lag behind because of the lack of a suitable transparent top electrode. High-performance semi-transparent plastic solar cells could be created by introducing an oxide-metal-oxide multilayer which was composed of MoO_3, with silver as a transparent top electrode[52]. The conductivity of the multilayer oxide electrode was governed by the intermediate silver layer sandwiched between two MoO_3 layers, and so the power-conversion efficiency also depended strongly upon the thickness of the intermediate silver layer in the oxide|metal|oxide electrode. By controlling the thickness of the silver layer, it was possible to obtain power-conversion efficiencies ranging from 4.5%, with some 50% maximum transparency in the visible region, to 9.1% with just 5% maximum transparency in the visible region. An ultra-thin gold capping layer on MoO_x reduced chemical modification and also acted as a transparent conductive electrode[53]. Such a layer could replace transparent conductive oxides such as indium tin oxide. The power conversion efficiency of a $Au|MoO_x|n$-Si device could be increased from 0.53 to 6.43% by incorporating a grid-type electrode into the front surface. The efficiency of a silicon solar cell was 4.2% with a 12nm-thick gold film, and was negligible (0.0008%) without the film. The further addition of gold grids could improve the conductivity of the ultra-thin gold capping layer and improve the power conversion efficiency. The sheet resistance decreased from $11163\Omega/sq$, for MoO_x/n-Si, to $9351\Omega/sq$ upon depositing a 2nm-thick gold film. The sheet resistance decreased markedly when the gold film thickness was greater than 4nm, with values of 182.72, 17.869 and $8.343\Omega/sq$ for thicknesses of 4, 8 and 12nm, respectively. The 2 and 4nm films exhibited discontinuous gold islands which were due to Volmer-Weber growth, in the initial stages, which was caused by a surface-energy mismatch. The discontinuous islands led to poor electrical conductivity and a high sheet-resistance. With increasing gold thickness, the separate clusters grew and merged into a continuous layer that imparted a good conductivity. There was a markedly reduced sheet resistance when the gold thickness was greater than 8nm. Upon fixing the MoO_x thickness at 10nm, the reflectance gradually increased with increasing gold thickness. The solar-weighted reflectance of a $Au|MoO_x|glass$ substrate was 12.7, 13.4, 15.8 and 20.5% for a gold film thickness of 2, 4, 8 and 12nm, respectively. The reflectance was relatively low at shorter (400 to 600nm) wavelengths, with a higher solar irradiance than that at above 800nm. The optical transmission was greatly affected by the gold film thickness. It gradually decreased with increasing thickness when the thickness was greater than 2nm. The transmission of 2nm gold films was different to that of thicker films, due to localized surface plasmon resonances of discontinuous gold islands. The calculated solar-weighted transmission values were 67 and 70% for 2 and 4nm films, respectively, on $MoO_x|glass$. Due to increased optical absorption arising from plasmon resonance, the 2nm film had a lower solar-weighted transmission than 4nm film, despite the lower thickness.

Table 6. Properties of transparent gold-capped MoO$_x$|n-Si solar cells

Gold(nm)	J$_{sc}$(mA/cm^2)	V$_{oc}$ (mV)	Fill-Factor(%)	Efficiency(%)
0	0.0096	501	16.5	0.0008
2	0.019	564	22.1	0.002
4	3.79	540	26.1	0.53
8	19.21	526	33.1	3.34
12	16.6	514	49.23	4.2

A discontinuous film led to a lower electrical conductivity. In this sense a discontinuous gold capping layer could be detrimental to device performance. In order to create ultra-thin continuous metal films, adhesion to the substrate has to be improved by reducing the interfacial free energy and surface energy. The insertion of a seed layer, or the use of a surface pre-treatment, could help to avoid a discontinuous metal film by preventing 3-dimensional island growth. The short-circuit current density varied widely with film thickness (table 6) for Au|MoO$_x$|n-Si solar cells. The open-circuit voltage was less affected by the film thickness (table 7). The highest value was found when the film thickness was 2nm, and gradually decreased with increasing thickness. The highest value, for 2nm, was attributed to a non-degraded MoO$_x$ hole-selective layer arising from air exposure. The fill-factor of Au|MoO$_x$|n-Si solar cells increased with increasing gold film thickness. The fill-factors were 16.5, 22.1, 26.1, 33.1 and 49.23% for film thicknesses of 0, 2, 4, 8 and 12nm, respectively. The improvement was attributed to the reduced sheet resistance of thicker films. The performance of the Au|MoO$_x$|n-Si solar cells was greatly dependent upon the gold film thickness. The efficiency of the cell was 0.0008, 0.002, 0.53, 3.34 and 4.2% when the film thickness was 0, 2, 4, 8 and 12nm, respectively. The highest efficiency (4.2%) for 12nm film was attributed to its much higher fill-factor. The fill-factor increased slightly with increasing MoO$_x$. The fill-factors were 44.07, 49.81, 49.23, 50.08 and 50.22% for MoO$_x$ thicknesses of 2, 5, 10, 15 and 20nm, respectively.

Table 7. Properties of Au|MoO$_x$|n-Si solar cells as a function of MoO$_x$ thickness

MoO$_x$(nm)	J$_{sc}$(mA/cm^2)	V$_{oc}$(mV)	Fill-Factor(%)	Efficiency(%)
2	18.69	446	44.07	3.67
5	17.55	531	49.81	4.64
10	16.61	514	49.23	4.2
15	15.93	523	50.08	4.17
20	15.55	509	50.22	3.98

By using thin metal films on optically transparent substrates, the interface can be accessed with infra-red light via a sufficiently thin electrode[54]. Platinum films with a thickness of about 200nm were deposited onto Al$_2$O$_3$(00•1) substrates by inductively-coupled plasma-assisted radio-frequency sputtering and annealed (700C), with or without oxygen. In the absence of oxygen introduction during sputtering, a (111)-oriented polycrystalline film was produced. The terrace widths were about 100nm and the film was of poor crystalline quality. The introduction of oxygen produced twinned monocrystalline epitaxial (111) films with a typical face-centred cubic surface morphology with facets of hexagonal symmetry. The root-mean-square roughness was about 0.4nm over an area of 8µm^2 and was less than 0.2nm over an area of 1µm^2.

Techniques

Oxidation-induced clustering and layering of gold, due to strong oxygen interference at the outermost surfaces of nanoscopic gold geometries was exploited in order to promote the evolution of gold clusters and layers that strongly wetted oxide supports[55]. Thus a 4nm-thick epitaxial gold layer could then exhibit both a higher optical transparency than that of silver, a near-bulk resistivity of 8 x 10^{-8}Ωm and an extreme resistance to chemical corrosion and mechanical deformation. This process provided a solution to the problem of transparent metal electrodes that are too vulnerable to degradation in work-environments. The details of the process involved a manipulation of the gold growth-mode so as to alter clustering and layering, with increased wetting occurring during the oxygen plasma-assisted gold deposition. The results confirmed that atomic-oxygen mediated clustering depended upon the segregation and incorporation of oxygen atoms at the topmost surface of evolving gold clusters. Oxygen-segregation at the surface led to a reduction in the free energy of gold clusters, and provided clear evidence for the surfactant behaviour of immiscible non-residual oxygen atoms during cluster evolution by suppressing the migration of gold atoms. The method could be applied to the creation of gold geometries ranging from clusters to a continuous layer. These were very different to the normal structural features which are generated by conventional vacuum-deposition. It markedly reduced the 3-dimensional growth of gold and greatly improved the crystallographic characteristics, due to defect annihilation at low temperatures.

A simple large-scale patterning technique for metal mesh was proposed[56] which involved the formation of a template on a transparent substrate by spray-coating, ink-jet printing, metal deposition and subsequent lift-off. An interdigitated electrode with a finger-gap of about 300μm was created over a large (30cm x 30cm) area. The patterning technique was further used to create an interdigitated electrode on a flexible poly(ethylene terephthalate) substrate. The transmittance spectrum of the patterned electrode revealed a transparency of about 91% at 550nm. A sheet-resistance of about 7Ω/sq was found for a deposited metal thickness of about 250nm. The opto-electronic properties were attributed to large transparent polygons between the highly interconnected narrow metal wires. The device had an intrinsic capacitance of about 7.8pF at 8MHz.

Core-shell silver/oxide particles (table 8) were prepared by coating silver onto the surface of SiO_2, ZnO or TiO_2 particles by using an alkylamine silver complex as the silver source[57]. The $Ag-SiO_2$, Ag-ZnO or $Ag-TiO_2$ nano-particles were then used to make a silver-metal oxide paste with epoxy resin as a binder, ethylcarbitol acetate as a solvent and a polymeric wetting agent as a rheological additive. A micron-sized trench pattern was produced on polyethylene terephthalate film via ultra-violet imprinting, using a mould with an embossed trench pattern. The nano-size silver paste was then used to fill the trench pattern. The resultant silver-mesh transparent conducting electrodes exhibited a visible-light transmittance of about 90% and a resistivity of $3.6 \times 10^{-4}\Omega$cm.

Semi-transparent mesh electrodes were prepared by means of the laser-structuring of a polymer, followed by the thermal evaporation of a metallic layer[58]. There was a reduction in ultra-violet laser line irradiation from 1.3W to about 0.3W, and the resist could be removed by water. With ethyl cellulose as the polymer, purpurin was used as a sensitizer which absorbed in the ultra-violet range; thus permitting structuring with a 355nm laser beam. This led to path-width reduction and better spatial resolution.

Table 8 Average particle-size and silver content
of silver-metal oxide nano-particles

Oxide	Particle Size(nm)	Silver Content(%)
SiO_2	120.8	40.4
SiO_2	101.4	37.6
SiO_2	98.7	36.3
SiO_2	96.6	34.8
SiO_2	112.3	38.7
TiO_2	151.6	37.6
TiO_2	122.0	35.6
TiO_2	90.7	34.7
TiO_2	120.4	33.5
TiO_2	100.7	41.3
ZnO	148.4	43.7
ZnO	98.7	33.6
ZnO	150.3	36.5
ZnO	121.7	34.6
ZnO	87.3	38.5

In the case of poly(vinylpyrrolidone), carbon black was added and the heat generated by the laser was exploited. The additive, antimony tin oxide, sensitized the polymer layer to infra-red and thus enabled structuring using the fundamental Nd:YAG laser-wavelength of 1064nm. Contact-less electro-reflectance and thermo-reflectance measurements were made of manufactured transparent electrodes. In the former case, two different metal-mesh electrodes gave the same result, while the thermo-reflectance measurement predicted divergent results. This was attributed to a difference in thermal modulation, depending upon the sample-heater configuration. Various shapes, with sharp or rounded edges, could be produced, and re-sizing the shape was easier and faster than in the case of traditional lithography. The method could be applied to the creation of flexible polymeric substrates. A metallic material with an adhesive layer could be deposited on the prepared disposable polymer mask. The materials which could be used were gold, silver, aluminium and copper. Titanium could be deposited as an initial 10nm adhesive layer. The thickness of the metal layer was controlled via the deposition time and was adjusted so as to attain a target value that ranged from 170 to 840nm. It was selected such that the material would be as continuous as possible over the entire surface could still be lifted off and detached after dissolving the polymer beneath.

A metal ink was formulated which could be sintered by means of intense pulsed light without any void-generation[59]. The ink was of hybrid type in which silver nano-particle ink and copper precursor ink were mixed in a weight-ratio such that the reduced copper filled the space between silver nano-particles during sintering. The formulation required optimization with regard to the shape and size of the nano-particles, the type of solvent and the proportion of copper precursor ink. Ethyl cellulose and 2-butoxyethanol were firstly mixed in the weight-ratio of 1:30, together with a wetting-agent. The silver nano-particle ink was prepared by adding silver nano-particles which weighed four times the weight of the mixed solution, and dispersed by using a 3-roll mill. The silver nano-particles were spherical, with a diameter of 50 and 100nm, and flake-like with a size of 500nm. The copper nano-particles were spherical, with a diameter of 100nm. The copper precursor ink was prepared by mixing $Cu(NO_3)_2 \cdot 3H_2O$ with ethyl cellulose and 2-butoxyethanol such that the concentration was 46wt%. Silver nano-particles were then added and dispersed in order to produce the hybrid ink. The total weight of $Cu(NO_3)_2 \cdot 3H_2O$ and silver nano-particles was 4 times that of the mixed solution. The weight-ratio of $Cu(NO_3)_2 \cdot 3H_2O$ and silver nano-particles was chosen to be 1:9, 2:8, 3:7 or 4:6 while the total weight was constant. In order to create metal-mesh electrode films, ultra-violet curing resin was applied to 100μm polyethylene terephthalate film, and the resin was then imprinted with intaglio patterns having a width of 4μm. The metal ink was filled into the trench patterns by using a blade. The metal-mesh electrode was then pulsed-light sintered by using a xenon lamp and visible light in the range of 320 to 1200nm. The pulse-duration was 20ms and energy ranged from 4.5 to 10.5 J/cm². This produced flexible transparent metal-mesh electrodes. When there were voids in metal lines following sintering, the lines could often shrink when exposed to high temperatures and high humidities for extended periods. This could impair the stability of the metal electrode. The present ink hardly shrank, due to an absence of voids in the metal lines following sintering, and possessed good environmental and mechanical stabilities. The sheet-resistance of the optimum metal-mesh electrodes did not increase by more than 2.1% after 30 days at 85C and 85% humidity. They were stable following 200 cycles of rapid thermal shocking between -45 and 125C. The metal-mesh electrodes which were produced by using the present ink were superior to transparent electrodes such as indium tin oxide or silver nano-wires when employed for heating or electromagnetic shielding.

A simple method was developed[60] which could form high-resolution patterns for the creation of transparent metal-grid conductors. It involved the embedding of conductive ink into a grooved structure that was formed by nano-imprinting. Capillary forces caused conductive ink to fill grooves with widths in the micron and sub-micron range. An aspect-ratio of 3.1 was obtained for a 1.6μm embedded pattern. A transparent conductive film was prepared which offered a transmittance of 82.7% and a sheet-resistance of 5.1Ω/sq at a grid-width of 3.0μm and a grid-pitch of 150μm. The grid area was 51mm x 51mm. The maximum figure-of-merit was 727. Even when the bending radius was just 2.5mm, the change in sheet-resistance was 27% and defects such as disconnections or short-circuits were absent. The grid was almost invisible to the naked eye. At a wavelength of 550nm, the transmittance of the bare film before processing was 91.3%. A critical bending radius of about 2.5mm was identified for the printed samples, while the critical radius of flexible indium tin oxide films ranged from about 8 to 10mm.

Transparent platinum electrodes were prepared via the nano-imprinting lithography of a molecular platinum complex, $Pt(NO_3)_2(ACN)_2$. The latter's thermal decomposition at temperatures below 220C, some 180C lower than that of $PtCl_2(ACN)_2$, produced line and grid patterns of elemental platinum[61]. Line patterns with a line-width of 200nm and a period of 1000nm, or of 0.61μm with a period of 2μm, were prepared on glass substrates. In both cases, homogeneous patterns were obtained which had a resistance as low as 95Ω over an imprinted area of 1cm x 1cm. The electrodes offered transmittances of 90%, and resistances which were less than 100Ω (table 9). This combination of precursor development and processing techniques was expected to permit the preparation of transparent highly-conductive electrodes which were based upon metals. The present electrodes had specific resistances which were very close to that of the bulk material, being as low as $2.2 \cdot x \ 10^{-7} \Omega ms$. For the first time, a resistance of less than 100Ω could be combined with a transmittance of up to 90%, over the imprinted area. The specific resistance of a 2μm-period pattern was calculated to be $4.3 \times 10^{-7} \Omega m$. As compared with platinum-line patterns having similar dimensions, there was an increase by a factor of 6 in the specific resistance. This improvement was attributed to a higher degree of crystallinity. Most of a decrease in transmittance was due to additional reflectance at the structured surface. High-quality line-patterns with a width of 0.38μm and period of 1.91μm were produced on transparent polyethylene naphthalate substrates which had a transmittance which was about 80% of that of the substrate. The resistance over the imprinted area was as low as 100Ω for small grids and 120Ω for large grids. For a given pattern geometry, the specific resistance of the material used has a marked effect upon

Materials Research Forum LLC
https://doi.org/10.21741/9781644903476

the sheet-resistance of the printed electrode. The transmittance increased with decreasing line-width, from about 68% for a large grid to 90% for a small grid. It was constant over the range of 400 to 900nm. For a 51.3nm silver grid and a window-width of 1000nm, a transmittance of about 85% was calculated; in good agreement with measured values. A key advantage of the novel precursor was an almost complete displacement of ink from the window area during printing. This produced highly transparent structured electrodes.

Table 9. Properties of transparent conductive platinum electrodes on a glass substrate

Pattern	Size	Transmittance(%)*	Resistance(Ω)
line	200nm	79	>1000
line	610nm	66	95
grid	small	89	100
grid	large	68	120

*average between 400 and 800nm

Stretchable transparent metal-grid electrodes have been described in which liquid gallium-indium (24.5wt%) eutectic alloy was used as the stretchable conducting component[62]. Photolithographic roll-painting and lift-off techniques created grids having a line-width of 20µm and a line pitch of 400 to 1000µm. These offered a transmittance of 75 to 88%, together with a sheet-resistance of less than 2.3Ω/sq. When powered by the freely deformable liquid-phase conductor, the grid maintained a stable conductivity during stretching by up to 40%. By combining the grids with an homogeneous transparent poly(3,4-thylenedioxythiophene):poly(styrensulfonate) layer, it was possible to construct highly stretchable uniformly conducting transparent electrodes. In order to judge the stretchability of the alloy grids, measurements were made of their sheet resistance during increasing tensile strain. The resistance gradually increased with strain but electrical connectivity was maintained, with no sudden increase in resistance or fracture occurring up to 60% of deformation. The increase in sheet resistance was caused by the increased length and reduced cross-sectional area of the stretched liquid-metal line and amounted to a factor of 2.56 after 60% stretching of a single interconnect. The resistance of the grid increased linearly by a factor of only about 1.5 at 60% strain. This reduced effect of stretching was partially attributed to contact resistances which accounted for an appreciable fraction of the total resistance but was not affected by stretching. Due to the Poisson effect, the cross-sectional area of alloy lines perpendicular to the stretching direction increased as the sample shrank in the direction normal to that direction. This compensated for the increase in resistance in lines parallel to the stretching direction. The fractional change in the sheet-resistance for various grid-pitches was very consistent, indicating that the alloy grids deformed in a similar manner regardless of their geometrical features. The sheet-resistance of an alloy grid (table 10) still returned to its initial value after 1000 cycles of 40% stretching, indicating that the grid was very resistant to the fatigue-failure which was common in solid electrodes. In networks of metallic nano-wires, an increase of more than 100% in the resistance during cyclic stretching had always been unavoidable because irreversible deformation or sliding of the nano-wires within an elastomer led to degradation of their electrical connections. In the present alloy structures, where the entire grid was a deformable liquid with no physical junctions, the sheet-resistance remained essentially constant; with a variation of less than a few percent following a large number of stretching cycles.

Table 10. Sheet-resistance of gallium-indium grids

Substrate	Line-Pitch(μm)	Resistance(Ω/sq)
rigid	400	1.25
rigid	600	1.69
rigid	800	1.88
rigid	1000	2.26
polydimethylsiloxane	400	2.25
polydimethylsiloxane	600	2.31
polydimethylsiloxane	800	2.76
polydimethylsiloxane	1000	3.13

An inkjet-printed indium tin oxide free electrode was made from particle-free silver ink, and an argon plasma was then used to reduce the ions in the ink to metallic silver[63]. The printed silver layers offered a good optical transmittance and electrical conductivity. Inverted indium tin oxide organic light-emitting diodes were produced by solution processing. The devices which contained the printed electrodes exhibited an improved luminance of 75% over the visible range and an improved current efficacy when compared with indium tin oxide based reference devices (table 8). The printed devices exhibited a high bending stability (figure 5) with regard to conductivity. The devices could also provide luminances of up to 12000cd/m^2. There was a high degree of flexibility when compared with that of indium tin oxide, and the resistance of the latter increased sharply after a few bending cycles. Metal films possess the highest conductivity at room temperature, but acceptable optical transmittance can be achieved only in the case of ultra-thin films[64]. Structuring the metal into the form of optically invisible nano-wires is a promising means for complementing or replacing transparent conductive oxides as transparent electrode material. The out-of-plane capability of electrohydrodynamic NanoDrip printing was used to pattern gold and silver nano-grids having line-widths ranging from 80 to 500nm. Its additive nature permitted the printing of high aspect-ratio nano-walls which appreciably improved the electrical behaviour while retaining a high-level optical transmittance. Metal-grid transparent electrodes were optimized to have a sheet-resistance of 8Ω/sq and a relative transmittance of 94%, or a transmittance of 97% and a sheet-resistance of 20Ω/sq. Silver-grid transparent electrodes were prepared[65] via electrohydrodynamic jet printing using silver nano-particle inks. A silver grid-width of less than 10μm was achieved by jet-printing, and was invisible to the naked eye. The silver-grid line-to-line spacing was varied in order to adjust the electrical and optical properties of the transparent electrodes. A decrease in the sheet-resistance at the expense of transmittance was observed as the silver-grid pitch was decreased. The pitch was chosen so as to optimize the electrical and optical properties. With a 150μm pitch, the jet-printed silver-grid electrode offered a sheet-resistance of 4.87Ω/sq and a transmittance of 81.75% in the near-infrared after annealing at 200C. A silver filling-factor was defined in order to predict the electrical and optical properties, and the measured electrical and optical properties were well-described by theoretical equations which incorporated this filling-factor. A gold-wire network which was nearly invisible to the naked eye was created[66] on substrates such as glass for use as a transparent conducting electrode. This involved coating a TiO$_2$ nano-particle dispersion to a thickness of 10μm, which then formed a crackle network which served as a sacrificial template for metal deposition. The resultant transparent electrode exhibited a visible transmittance of 82% and sheet-resistance of 3 to 6Ω/sq, with a metal fill-factor of 7.5%. Sub-micrometre transparent silver and gold network electrodes which were invisible to the naked eye were incorporated[67] into organic photovoltaic devices. The spontaneous cracking of a polymer layer was used as a template for the preparation of the metal network. The silver or gold electrodes offered a transmittance of 80% over the visible

Materials Research Forum LLC
https://doi.org/10.21741/9781644903476

range. The figure-of-merit for transparent electrode materials had long been defined[68] to be T/R, where T was the optical transmission and R was the sheet-resistance. Expressions were originally derived for predicting the transparent electrode properties of a material on the basis of its fundamental electrical and optical constants. These electrodes also exhibited an ultra-low haze of about 5 % and had a high figure-of-merit. An indium tin oxide-free semi-transparent polymer solar cell which incorporated silver or gold network electrodes exhibited 57 % transmittance above 650nm. Conductive transparent gold or silver nano-wire mesh films were prepared[69] during which the nano-wires were produced by the reduction of metal ions when a thin growth solution was spread on a substrate. The metal reduction occurred within a template on the substrate so as to form ordered bundles of ultra-thin nano-wires. The films exhibited metallic conductivity over large areas, together with high transparency and flexibility. An investigation was made[70] of the effect of geometrical lattice modifications upon the optical and electrical properties of silver-grid electrodes. A reference grid with a width of 5µm and a pitch of 100µm was prepared by means of photo-lithography and lift-off and exhibited a sheet-resistance of 13.27Ω/sq, a transmittance of 81.1% and a figure-of-merit of 129.05.

Table 11. Turn-on voltage, luminance and current efficacy of
organic light-emitting diodes

Electrode	Voltage(V)	Luminance(cd/m^2)
indium tin oxide	2.5	16636
silver	2.5	11983
flexible indium tin oxide	3.0	3890
flexible silver	2.5	2170

Modified grid electrodes with stripe-added mesh, triangle-added mesh and diagonal-added mesh were suggested to improve the optical and electrical properties. All of these modified electrodes offered transmittances of 72 to 77% and sheet-resistance of 6 to 8Ω/sq. All of the modified electrodes exhibited appreciable improvements in the figure-of-merit, with the highest value being 171.14 for the diagonal-added mesh grid. This was comparable to that, 175.46, of indium tin oxide electrodes. The feasibility of a diamond-added silver-grid electrode for use in organic solar cells was confirmed by means of finite-difference time-domain simulations. Unlike the oxide electrode, the diamond-added electrode could induce light-scattering and trapping due to a diffuse transmission that compensated the loss in optical transparency.

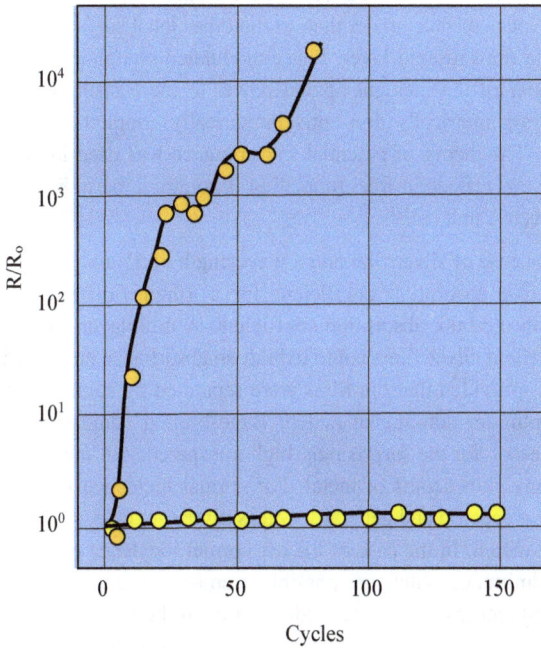

Figure 5. Resistivity change due to repeated bending of printed silver and indium tin oxide electrodes on polyethylene terephthalate. Orange: indium tin oxide, yellow: printed silver

By placing a composite layer, consisting of dielectric and metallic strips, on top of a metallic one, back-scattering from the metallic film could be almost exactly cancelled by the composite layer under certain conditions, leading to transparency of the entire structure[71]. Experiments which were performed in the terahertz domain, using metamaterials to mimic plasmonic metals in the optical regime, yielded good agreement with theory. An investigation of the conditions required for the complete tunnelling of light through a composite barrier comprising impedance-mismatched metamaterial layers showed that two types of complete tunnelling were possible[72]: phase-unmodulated and

Materials Research Forum LLC

https://doi.org/10.21741/9781644903476

phase-modulated complete tunnelling. Local surface modes which formed near to the interfaces between the metamaterial barrier layers were pivotal in complete tunnelling.

Phase-unmodulated complete tunnelling occurred via successive mode-couplings: from incident light to the local surface mode, then to the other local surface mode and then to the light-mode in the transmission layer. Phase-modulated complete tunnelling resulted from complete transfer of the incident optical power to the transmission layer, via the direct mediation of symmetrically and antisymmetrically coupled super-modes of the local surface modes. The theory of potential transmittance was used in order to derive a general expression for reflection-free tunnelling through a periodic stack having a dielectric-metal-dielectric unit cell[73].

As an example, in the case of silver film and a wavelength of 550nm at normal incidence, the minimum effective absorption coefficient for a 10nm film was two orders-of-magnitude lower than the bulk absorption coefficient. A multilayer which contained an arbitrary number of 10nm silver films could exhibit an absorbance embodied by the same minimum coefficient provided that the films were separated by suitable dielectric layers for producing an optimum admittance match. Band-limited admittance matching was supposed to be the reason for the surprisingly high transparency of metal|dielectric stacks which contained many skin depths of metal. For normal incidence from the air side, it was shown that only a specific and generally unfeasibly large dielectric index will permit a perfect admittance-match. In the case of the off-normal incidence of transverse electric polarized light, an admittance match was possible at an isolated angle that depended upon the indices of the ambient and dielectric media and upon the thickness and index of the metal. For transverse magnetic polarized light, admittance-matching was possible within the evanescent-wave range where tunnelling was mediated by surface plasmons. In all of the cases, matching was predicted to occur at a single wavelength and/or tunnelling angle. The use of two or more dielectrics was expected to create structures that were matched at more than one wavelength or tunnelling angle. It has been shown that silver could, due to an analogy with tunnelling, be rendered completely transparent by using impedance-matching media in front of, and behind, the metal[74]. Optical experiments exploited beyond-the-critical-angle barriers in a frustrated total internal reflection arrangement which mimicked quantum mechanical systems. The same mechanism also permitted greatly increased transmission through unstructured thin metal films without requiring surface wave excitation. The thin silver film was completely opaque, but it transmitted light when it was sandwiched between transparent layers. This optical effect was mathematically identical to the quantum-tunnelling of a particle through a barrier. Neither process normally permits complete transmission, but a more complicated gap or barrier structure could lead to efficient transmission. An extra layer of a different transparent material was added before and after the air-gap. At certain wavelengths, perfect transmission could occur via the cancelling-out of light-waves reflecting between the sandwich layers. One face of each of two glass prisms was coated with some 200nm of zinc sulphide. The coated faces were then pressed together to leave a thin air-gap. Laser light which hit a prism at a suitable angle then penetrated the sulphide|air|sulphide sandwich with some 85% efficiency. When the air-gap was replaced by a 40nm metal film, light was still transmitted with an efficiency of up to 35% at certain wavelengths.

On the basis of theoretical analysis and full-wave simulations[75], a method was proposed in which a continuous metal film could be rendered transparent in the optical regime, with light scattered from the metal film being cancelled-out by that from two composite layers which consisted of metallic and dielectric stripes. The transparency and conductance of the metal were retained and the method was insensitive to incidence-angle and structural problems.

A continuous highly conducting metal film could be made transparent to the wide-angle and polarization-insensitive incidence of near-infrared light by depositing periodic metal patches on top of the metal film[76]. This system could suppress reflection and enhance transmission. The theory of potential transmittance was used[77] to derive a general expression for reflection-less tunnelling through a periodic stack which had a dielectric|metal|dielectric unit cell. In the case of normal incidence from air, theory showed that only an impractically large dielectric index could generally lead to a perfect admittance match. In the case of the off-normal incidence of transverse electric polarized light, an admittance match was possible at a particular angle which depended upon the indices of the ambient and dielectric media and upon the thickness and index of the metal. In the case of transverse magnetic polarized light, admittance matching was possible within the evanescent-wave range. That is, for tunnelling which was mediated by surface plasmons. An early study concerned the optical properties of nanoscopic metal particles and the fact that bulk metals are optically opaque while effective medium theory predicts that collections of small metal particles can be transparent if the wavelength of light which is involved is much longer than the characteristic dimension of the particle[78]. Effective medium theory could be used to identify metal-containing composites that were electronically conductive and optically transparent. Composites of this type could be prepared by means of the electrochemical deposition of metals within the pores of a micro-porous so-called template membrane. It was shown that composite membranes which were prepared in this way could be optically transparent over the entire near-infrared range, and into the visible region. By changing the shapes of the particles, the colour of the composite could be varied. Arrays of nanoscopic gold cylinders were prepared via the electrochemical deposition of gold within the 50nm pores of nano-porous alumina template membranes, and these could be optically transparent into the visible range. By changing the aspect-ratio of the gold nano-cylinders, the colour of the composite membrane could be varied. The results were in accord with the predictions of the Maxwell-Garnett theory of metal-insulator composites

Negative refraction of the Poynting vector, and sub-wavelength focusing within the visible range, were demonstrated by using a transparent multilayered metal-dielectric photonic band-gap structure[79]. Within the wavelength-range of interest, evanescent waves were not transmitted, and the main fundamental physical mechanisms of sub-wavelength focusing were resonance-tunnelling, field-localization and propagation-effects. The structures offered tunability and a transmittance of at least 50% over the visible and near-infrared ranges. The results suggested a simpler means for preparing a material with a negative refraction and a high transparency over a broad wavelength-range. Although bulk silicon and GaAs are opaque within the visible range, several relatively thin but resonant layers could be combined, with various thicknesses of silver so as to produce a range of transparency. A chirped 13-layer stack which comprised Ag(32nm)/Si(22nm), in which the first and last layers were 11nm thick, offered a transmittance window which centred around 500nm, with a maximum transmittance of some 20%. At this wavelength a negative refraction angle of about 10° was anticipated. Although GaP has a lower index of refraction than does silicon, largely similar results were obtained by creating a stack which contained almost 260nm of silver. At 632nm, the transmittance of a chirped stack of Cu(32nm)/Si(43nm) was about 30%, with a negative refraction angle of about 3°. It was thus possible to use common materials with transparent metal stacks, and still retain negative refraction and sub-wavelength resolution within the visible range. These structures could contain hundreds of nanometres of metal, but be transparent to visible light. Simulations showed that incident transverse magnetic polarized waves underwent negative refraction. These seminal results offered a new point of view regarding light-propagation in metals. Because transparent metal structures had been fabricated for the visible part of the spectrum, by using various dielectrics, the experimental production of negative refraction in such structures was not expected to encounter new technical hurdles.

Metamaterials have a surprisingly long history, with the first one being described in 1948[80], when a metallic lens antenna was described in which focusing was achieved by reducing the phase-velocity of radio waves[81]. The reduced velocity was due to the presence of conducting elements whose length in the direction of the electric vector of the impressed field was small in comparison with the wavelength. They acted as small dipoles which were analogous to the molecular dipoles which are set up in non-polar dielectrics by an impressed field. The lens then exhibited the broad-band characteristics

Materials Research Forum LLC

https://doi.org/10.21741/9781644903476

of a solid dielectric lens. With regard to metamaterials such as transparent metals a scattering-cancellation mechanism was proposed for making continuous metals optically transparent. A possible scheme was proposed, and was proved for the microwave and terahertz frequency-domains. Light reflection and refraction at a suitably-designed gradient-index meta-surface was shown to obey a generalized form of Snell's law. The system could perfectly convert a propagating electromagnetic wave to a surface wave under some conditions. An optically transparent antenna was based[82] upon a micro-metal mesh conductive film. It consisted of two meandering monopole radiators. The film was constructed by using self-assembling nano-particle technology and offered a transmittance of above 75% together with a sheet-resistance of 0.05Ω/sq. One-dimensional photonic band-gap structures have also been termed transparent metals, and have been tested as electromagnetic shielding in the radio-frequency range by using physical vapour deposition[83]. Transparent electromagnetic shields, based upon metal-dielectric multilayered coatings, have been applied to glass and plastic substrates.

Colloidal crystals with diameters of 200 to 500nm were found to be sedimented vertically along the [100] direction of the face-centred cubic lattice[84]. Various metallic and dielectric photonic crystals were fabricated from templates which were composed of spheres with diameters ranging from 200 to 2300nm. Silver colloids which were 250 to 1200nm in diameter were created which had a uniform size distribution and were assembled into opals. Conducting polymers and gold were then infiltrated into the interstices of the synthetic opals so as to use the crystalline structure as a template. The spheres were etched away to leave inverse opal photonic crystals. Control of the shape and formation of such particles was difficult and only a few geometries with uniform populations of colloids could be produced. Colloidal gold discs with a diameter of 4μm and a thickness of 1.5μm were created by using red blood cells as templates. Control of the osmotic pressure permitted control of the cell-shape and led to uniform populations with various shapes, including spheres. Gold disc particles were used to build composite colloidal particles which consisted of alternating layers, and were assembled into small colloidal crystals. Colloidal silica spheres having diameters of 200, 250 and 290nm were self-assembled with monocrystalline crystallites which were 4 to 5mm wide and 10 to 15mm long. Larger spheres, with diameters of between 1000 and 2300nm, were also self-assembled with monocrystalline crystallites which were up to 1.5mm wide and 2mm long. The silica opals self-assembled vertically along the [100] direction of the face-centred cubic lattice, resulting in self-templated opals. Inverse opal photonic crystals with a partial band-gap with a maximum in the near-infrared were constructed from opal templates that comprised 2300nm spheres with $Ge_{33}As_{12}Se_{55}$; a transparent glass in the near-infrared with a high refractive index. Inverse gold and gold|polypropylene composite photonic crystals were fabricated from synthetic opal templates which were composed of 200 to 290nm silica spheres. Gold was infiltrated into the opal templates as gold chloride and was thermally converted into metallic gold. Opals which were partially infiltrated with gold were co-infiltrated with polypropylene for mechanical support, before removal of the silica. Lithography-patterned substrates served as guides which forced colloidal spheres to assemble into a face-centred cubic crystal lattice, vertically along the [100] direction. The point of all of these operations is to create a material possessing a periodically varying index of refraction. This is necessary in order to produce an optical band-gap which then leads to photonic band-gaps in the visible and near-infrared regions. The excellent control of sheet-resistance and transmittance which is offered by metal-mesh electrodes is a great advantage for micro-electronic applications, but weak adhesion between the mesh and a substrate has hindered its use in flexible opto-electronic devices. One approach[85] is to combine colloidal deposition and silver enhancement. Adhesion of the metal mesh is greatly improved by introducing an intermediate adhesion layer. Patterns which were fabricated with a minimum feature-size of 700nm led to a transmittance of 97.7% and a conductivity of $71.6\Omega/sq$. A transparent heater could be constructed from silver mesh which produced a temperature of up to 245C using 7V. Flexible semi-transparent heating arrays were prepared[86] by gray-scale printing. A perforated silver mesh with tunable features was first prepared by varying the toner gray-scale during laser printer. The resistance could be varied from 1 to $2\Omega/sq$ by changing the silver fill-factor from 50 to 70%. A silver heater based upon joule heating could produce temperatures of up to 135C for hours. Porous Si-SiO_2 micro-cavities have been used[87] to modulate a photo-detector having a detection range of 300 to 510nm. All of the micro-cavities had a localized mode which was close to 360nm, such that the porous Si-SiO_2 filters cut the photo-detection range from 300 to 350nm; where the micro-cavities offered low transmission. Visible-light blind and flexible metal–semiconductor–metal ultraviolet photo-detectors were based[88] upon less than 10nm thick silver interdigital electrodes. Transparent detectors with 7nm electrodes offered a transmittance of 80% and a responsivity of 60.5mA/W and a detectivity of 1.75×10^{10} Jones at 5V under 380nm illumination. It was as fast as the equivalent opaque detector, with rise and fall times of 22.4 and 11.5ms, and exhibited good flexibility. The improved properties were largely attributed to a broad high ultraviolet-to-visible transmittance and to the high conductivity of the ultra-thin silver film. An additional blocking component such as monolithic deposited silver had been used to remove unwanted out-of-band radiation, but this also reduced the overall transmission through the filter. Pairs of $Ag|SiO_2$ layers and a layer of silica were added as an antireflection coating, with the band-pass filter being designed to have maximum transmission in the UV-A range, with a resonance at 320nm.

The use of transparent metals ensured a transmission decrease of several orders of magnitude in the visible and infrared range, at the same time, while preserving transparency in the ultraviolet spectrum. Deep-UV band-pass filters which were based upon a mixture of aluminium and SiO_2 were also developed. Those filters had a 27% transmission peak at 290nm, a band-pass from 250 to 350nm and a rejection ratio to visible light of 20dB. The peak transmission could be tuned by adjusting the metal dimensions. A disadvantage of $Ag|SiO_2$ and $Al|SiO_2$ structures was that they led to a lower transmission of less than 50% in the ultraviolet range, due to absorption by the metals. In the short-wavelength range, photons were absorbed and did not contribute to the photocurrent. The density of recombination centres was thus very high and the photodetector-sensitivity was lower than it was without a filter. The maximum transmission between 356 and 364nm predominated in the UV-A range, and permitted the flow of high-energy photons. The filters favoured light transmission at 390 to 510nm, where photons contributed to the photocurrent. A solar-blind deep-ultraviolet pass filter was designed[89] which had a 27% transmission peak at 290nm, a pass-band width of 100nm (250 to 350nm) and a 20dB rejection ratio between deep-ultraviolet and visible wavelength. The filter consisted of an aluminium nano-grid which was made by coating 20nm of aluminium onto a silica square grid having a 190nm pitch, a 30nm linewidth and a 250nm depth. Its behaviour could be explained by a coupled wave analysis. The wavelength of peak transmission and the pass-band width could be tuned by adjusting the metal nano-grid dimensions. Paired media which have the same permittivity and permeability, but with opposite signs, can optically cancel out the other member of the pair. The creation of such complementary media tends to require metallic resonating structures. All-dielectric unidirectional complementary media were created by using symmetrical dielectric multilayers[90]. These could be treated as effective media, having flexible effective permittivity and permeability, which could function as unidirectional complementary media of almost any type, including metal. It was shown that unidirectional wave transmission through metal films could be greatly increased by using all-dielectric unidirectional complementary media. As a specific example, magnesium fluoride with a refractive index of 1.38 and titanium dioxide with a refractive index of 2.35 were used as the dielectrics. The effective parameters of the $MgF_2|TiO_2|MgF_2$ symmetrical dielectric multilayer were deduced and a silver film was used as an optically opaque wall. It was first assumed that the silver was lossless. From the effective parameters of the symmetrical dielectric multilayer it was found that it satisfied the equation,

$$\varepsilon_e/\varepsilon_w = \mu_e/\mu_w = d_w/D$$

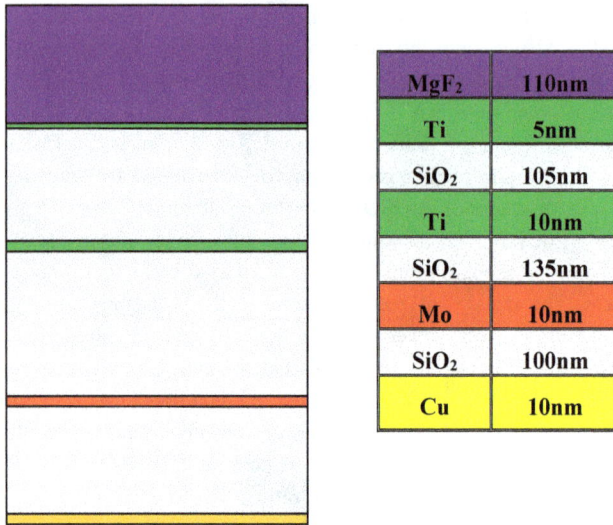

Figure 6. Side-view of a bidirectional metamaterial absorber

where ε is the permittivity, μ is the permeability and the subscripts indicate multilayer and wall, while d_w is the wall thickness and D is the total thickness of the symmetrical dielectric multilayer. When these conditions are satisfied, both reflection and phase accumulation within the wall can be eliminated by the symmetrical dielectric multilayer, due to multiple-reflection interference. As a result the overall dielectric|wall system can be considered to be a null space with regard to incident light, leading to perfect light transmission. This thus offered an efficient approach to transmission enhancement and light transmission through opaque walls under normal incidence was possible. By adjusting the thickness of the components, the symmetrical dielectric materials could function as effective unidirectional complementary media for almost any material. Light transmission through a silver film could be greatly enhanced even by using the dielectrics at a distance.

A study was made of the effect of various structural parameters and shapes upon the absorption of a bidirectional metamaterial (figure 6)[91]. Simulations showed that, when the thickness of an MgF_2 layer was fixed at 110nm, the absorption spectrum of the absorber had peaks at wavelengths of 430, 665, 880 and 150nm. Upon increasing the thickness of the fluoride layer from 90 to 120nm there was a corresponding change in the absorptivity, which exceeded 90.0%, from the 410 to 2220nm to the 415 to 2310nm range. The absorber exhibited a maximum and an average absorptivity of 99.9 and 97.0%, respectively, across an ultra-wideband range of 1900nm. It could exhibit a maximum absorptivity of 95.7% at a wavelength of 480nm upon changing the incident direction.

When gold, aluminium or nickel was used as the metallic component, the absorption peaks shifted to 495, 475 and 360nm, respectively. This resulted in corresponding decreases in the maximum absorptivity to 93.0, 91.3 and 67.5%. The exact choice of metal had essentially no effect upon the absorption spectrum. The distributions of electric and magnetic field intensities indicated that the magnetic fields which were associated with the various excitation wavelengths were distributed within the fluoride layer, the top titanium layer and the top SiO_2 dielectric layer. Resonance of the Fabry-Perot cavity within an absorber could be produced by incorporating two parallel flat semi-transparent metals which acted as the mirrors. Layers of titanium, molybdenum and copper were separated by dielectric layers consisting of loss-less SiO_2. The metal layers, separated by SiO_2 layers, constituted two distinct Fabry-Perot cavities. Excitation wavelengths of 430, 665, 880 and 1510nm satisfied cavity resonance conditions. The intensities of the magnetic field distributions, which originated from the cavity-resonances within the 3 SiO_2 cavity layers, varied depending upon the various excitation wavelengths. When normal incident light interacted with the absorber it underwent constructive or destructive interference with the reflected light, leading to distinct patterns of magnetic field distribution, and the latter were evidence of the occurrence of Fabry-Perot cavity resonance within the SiO_2 dielectric layers. When the excitation wavelengths were 430, 665, 880 and 1510nm, plasmon-enhanced surface plasmon resonance was excited in the fluoride, upper titanium and upper SiO_2 layers.

An absorber which offered high absorptivity over the visible to near-infrared range was analyzed[92]. The absorber (figure 7) was constructed from 2-layer cubes which were stacked on 4-layer continuous plane films. The 2-layer cubes comprised titanium dioxide and titanium and the 4-layer plane films consisted of poly(N-isopropylacrylamide), titanium, silica and titanium. The primary cause of the high absorptivity was the anti-reflection effect of the top TiO_2 layer. A secondary cause was that 3 different resonances, localized surface plasmon resonance, propagating surface plasmon resonance and Fabry-Perot cavity resonance, coexisted in the absorption peaks; at least two of which could be excited at the same time. A third cause was that two Fabry-Perot resonant cavities were formed within the poly(N-isopropylacrylamide) and silica dielectric layers. Because of the combined anti-reflection and resonance effects, the absorber offered high and wide-band absorptivity. Absorption peaks at 470, 510, 780, 940, 1280 and 2880nm were present in the absorption spectrum. The peak at 2500 to 3200nm exhibited an apparent red-shift from 2730 to 2950nm as the side-length of the cube was increased from 220 to 260nm. This proved that the peak was caused by propagating surface plasmon resonance.

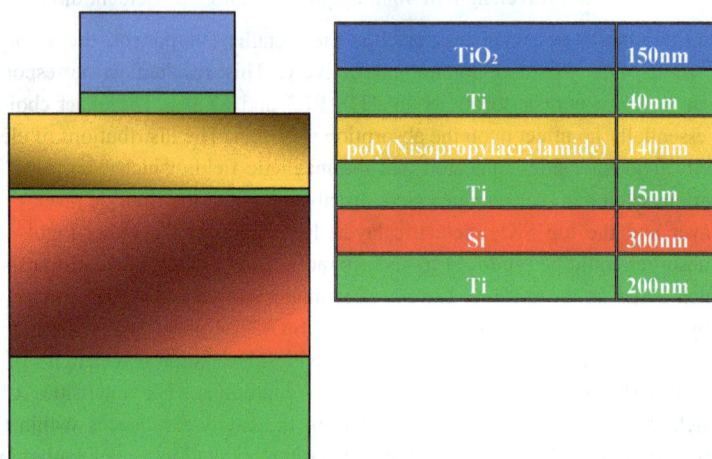

TiO$_2$	150nm
Ti	40nm
poly(Nisopropylacrylamide)	140nm
Ti	15nm
Si	300nm
Ti	200nm

Figure 7. Side view of the unit cell of the device

Numerical analysis showed that the absorber exhibited an absorptivity of more than 90%, and an average absorptivity of 97.9%, between 610 and 3600nm. The maximum absorptivity was 99.9%. Magnetic fields with differing exciting wavelengths were distributed in the TiO$_2$ and titanium cubes and in the poly(N-isopropylacrylamide) and silica dielectric layers. Fabry-Perot cavity resonance could be generated by two parallel flat semi-transparent metals acting as mirrors. The titanium layers were separated by loss-less poly(N-isopropylacrylamide) and silica dielectric layers and the titanium layers, separated by the poly(N-isopropylacrylamide) and silica layers formed two Fabry-Perot cavities. Because the exciting wavelengths of 470, 510, 780, 940 and 1280nm satisfied the conditions for Fabry-Perot cavity resonance, normally incident light underwent constructive or destructive interference with the reflected light. The magnetic-pattern distribution thus showed that the Fabry-Perot cavity resonance was actually excited in the poly(N-isopropylacrylamide) and silica layers. Because the exciting wavelengths were different, the intensities of the magnetic field distributions, caused by the Fabry-Perot cavity resonances in the layers of poly(N-isopropylacrylamide) and silica were different.

The magnetic-pattern distributions showed that because the exciting wavelengths were 470, 510, 780, 940 and 1280nm, localized surface resonance was also excited in the $TiO_2|Ti$ cubes. Perfect absorptivity around 780 and 940nm was attributed to a combination of the localized surface plasmon and Fabry-Perot cavity resonances. Because the wavelength of normally incident light was 2880nm, the electric field was distributed around the 2-layer cubes while the magnetic field was distributed over the entire absorber. Thus the perfect absorptivity at about 2880nm was mainly due to propagating surface plasmon resonance over the entire device.

A study was made of a self-assembly method for the construction of gradient-index lenses of extremely small dimensions[93]. By manipulating densely-packed arrays of metallic nano-particles it was possible to exert great control over the local refractive index. This was achieved via the layer-by-layer assembly of gold nano-particles, of various sizes, over silica beads. A gradient-index lens light-sink permitted light to be preferentially directed towards the centre, and allowed the creation of light-sinks smaller than 2.5μm. Developments in such metamaterials have demonstrated that compact metallic materials can form highly transparent metallic structures having controllable refractive indices for wavelengths within and beyond the near-infrared. High refractive indices are possible in the visible range. These metamaterials also lead to marked electromagnetic-field enhancement in the gaps between metal nano-particles. These so-called transparent metals generally consist of close-packed metal nano-particles and, due to their small dimensions, the material can behave as an effective dielectric. This behaviour was due to the limitation on the ability of the conduction electrons to travel and oscillate only within the nano-particles and behave like an atom in a dielectric. In order to confine electron movement, the nano-particle has to be isolated within an insulating layer. The effective refractive index of the structure can be obtained by solving the complex band-structure of the system. Films of gold nano-particles with diameters of 60, 33 and 15nm have been produced which offered a high transmittance in the near-infrared range and losses in the visible region. It had been noted[94] that, although metals are highly opaque, densely-packed arrays of metallic nano-particles can be more transparent to infrared radiation than are dielectrics such as germanium, even when the arrays comprise more than 75vol% of metal. Such arrays constituted effective dielectrics which were essentially dispersion-free over ultra-broad ranges of wavelength from microns up to millimetres. The local refractive index could also be tuned by altering the size, shape and spacing of the nano-particles. It had also been found[95] that assemblies of gold nano-particles, engineered using DNA, exhibited an ultra-low dispersion dielectric response in the infrared range. Broadband hyperspectral mapping of transmission and dark-field scattering revealed a polarization-insensitive optical response with distinct features in the visible and near-infrared ranges. Study identified a universal and characteristic response which was defined by a band of multipolar Mie resonances which depended only weakly upon the crystal size and light polarization. The use of gold superlattice microcrystals as scattering materials was thus established. It is expected that in a spherical system, as more layers of equal-sized nano-particles are assembled over a micro-bead, empty spaces are occupied by nano-particles. The dielectric-medium approximation for such a region then becomes more valid and creates a larger effective dielectric area within which light can travel and concentrate towards the inside, leading to a lens-like behaviour.

It was noted that metamaterials usually rely on the resonant behaviour of their constituent blocks and that this greatly limits their exploitation for handling certain frequency ranges. Some novel metamaterials were composed[96] of densely-packed metal nano-particles but behaved like dielectrics. They were highly transparent for all wavelengths within or beyond the near-infrared and their behaviour was constant across an ultra-broad range of frequencies. It was possible to tune the refractive index of these metamaterials to extraordinary high values while maintaining transparency. The real part of the effective refractive index of arrays involving different metals, but with the same geometry and surrounding refractive index, tended to the same value. The imaginary part of the effective refractive index however was different. This was attributed to differing skin-depths for the metals. It was suggested that metals with greater skin-depths produced the most transparent and least dispersive nano-particle arrays. The presence of surrounding material between the metal nano-particles led to appreciable amplification of the effective refractive index, as compared with metallic metamaterials in a vacuum. It was thus possible to obtain an extremely high effective refractive index by reducing the gap-width and filling the gap with materials having a higher refractive index. A single-layer anti-reflection coating could be created by using the simplest interference concept. This required a single transparent thin layer of material having a refractive index which was equal to the square root of the refractive index of the substrate. Gold nano-cylinder arrays could be used to form the anti-reflection coating layer. As the period of the array was 50nm, the layer thickness was 4μm. In order to make a system more transparent, metals such as titanium, with its greater skin-depth, could be used. It was expected that it would be easy to create devices that could operate over a broad range of frequencies by using metal dielectrics which exhibited a constant performance over the same frequency range. The transparency was not related to the generation of plasmon bands resulting from the structural periodicity and, at the visible wavelengths where surface plasmon resonances occurred, even metallic arrays with a low filling-fraction exhibited losses. At longer operating wavelengths, metallic particles within metal dielectrics, although being smaller than the metal skin-depth, exhibited negligible losses and the arrays were highly transparent.

The creation of good transparent conductors requires simultaneously increasing electrical conductivity and increasing optical transparency. Metals, with their high electrical conductivity unfortunately have high carrier densities and this pushes the plasma edge into the ultra-violet range. This can be countered by reducing the conductor thickness or the carrier density, at the cost of suffering lower conductance[97]. Highly anisotropic crystalline conductors offer another solution, by separating the directions of conduction and transmission. The ironically named so-called metal, Sr_2RuO_4, is optically transparent even at macroscopic thicknesses for c-axis polarized light. Creating a transparent 'metal' requires fine-tuning of materials parameters in the visible range. It is also necessary to ensure that interband transitions occur only above the ultra-violet so as to avoid absorption. Itinerant electrons meanwhile should react slowly in comparison to incoming light, so as to avoid screening and reflection. Pure metals such as copper, gold and palladium have high carrier densities and high conductivity and therefore have a plasma frequency in the blue to ultraviolet range and a metallic appearance. The simplest way to shift the plasma frequency to below the visible range is to use materials with a reduced carrier density. This then leads to a reduced electrical conductivity according to the Drude theory. Another approach is to reduce the plasma frequency via effective-mass tuning in electronically correlated quantum materials. Transparent conductors were obtained here by separating the conducting and transmitting directions in highly anisotropic metals. In order to demonstrate this generic behaviour, attention was focused on the 'metal' Sr_2RuO_4, which consisted of RuO layers separated by strontium. At room temperature the out-of-plane conductivity is reduced by more than two orders of magnitude with respect to the in-plane value. A 400nm slab of Sr_2RuO_4 was mounted on a sapphire substrate with pre-patterned aluminium contacts. The room-temperature resistivity was unaltered. At low temperatures, bulk Sr_2RuO_4 monocrystals had a resistivity of less than $1\mu\Omega$cm. The present device remained more resistive. The in-plane conductivity at room temperature was 1.25S/cm. The c-axis conductivity was 7mS/cm. The study of transparency was based upon separating the directions of high and low electrical conductivity via the crystalline anisotropy of Sr_2RuO_4. Linearly polarized light with an electric field in the in-plane direction was easily screened by highly mobile in-plane electrons and thus reflected. Out-of-plane electric fields were hardly screened and thus photons were transmitted over long distances. In order to satisfy the competing demands of high transmittance and conductivity, one strategy is to increase the carrier concentration, in a wide-bandgap semiconductor with low effective carrier mass, by means of heavy doping, as in the case of tin-doped indium tin oxide[98]. An alternative strategy has been suggested for creating high-conductivity high-transparency materials which relies upon strong electron–electron interactions, resulting in an increase in the carrier effective mass. The approach was verified experimentally for the so-called correlated metals, $SrVO_3$ and $CaVO_3$. Here, in spite of their carrier concentrations of more than 2.2 x $10^{22}/cm^3$, they possess screened plasma energies of less than 1.33 eV. There is also a desire to replace glass with so-called 'transparent aluminium' for certain applications. This is another ironic reference to the ceramic, $AlON$[99]. This has a density of $3.67g/cm^3$, a transparency of about 80% from the near-ultraviolet to the visible region and near-infrared, is solid up to about 1200C, is 3 times harder than steel and is scratch-resistant. One drawback is that the graphite dies which are used for hot-pressing can contaminate the sintered material at high temperatures.

A 2-step surface-energy directed-assembly process has been developed for the preparation of high-resolution silver mesh[100]. The method replaced assembly on a functionalized substrate, with hydrophilic mesh patterns, by assembly on a functionalized substrate with stripe patterns. During the process, a 3-phase contact line is pinned on the hydrophilic regions of the pattern but recedes on the non-patterned hydrophobic regions. The use of metal nano-meshes is a useful technique, but the development of low-cost electrodes is a possible bottleneck[101]. One method for producing highly flexible metal nano-mesh electrodes at a very low cost has been based upon applying a de-alloying process to ultra-thin Au-Cu alloy films which involved acid vapour. The nano-meshes could be transferred to any planar or curved support. By using this approach, it was easy to fabricate gold nano-mesh electrodes transferred onto polyethylene terephthalate films and thus offer 79% transmittance, with a sheet resistance of down to 44 Ω/sq; together with high stability under severe deformation. Silver meshes having a line-width of as little as 2μm can be assembled on either rigid or flexible substrates, and their thickness can be adjusted by varying the withdrawal rate and assembly time. The assembled meshes exhibit opto-electronic properties such as a sheet-resistance of 1.79Ω/sq, an optical transmittance of about 92% and a figure-of-merit of 2465; together with good mechanical stability (figure 8). Following assembly of the silver mesh, annealing is required in order to sinter and merge the nano-particles and ensure a high conductivity. In a typical case the process was used to produce a pattern which was 40μm by 100μm by using a withdrawal rate 3000μm/s for various times, yielding a thickness of about 150nm. At an annealing temperature of only 120C, no coalescence of the nano-particles occurred and isolated nano-particles remained. With increasing annealing temperature, coalescence occurred and correctly sintered films were formed. The resistivities of the annealed patterns were measured using 4-probe methods. As the annealing temperature was increased from 100 to 250C, the resistivity decreased by some two orders of magnitude, due coalescence of the nano-particles (figure 9).

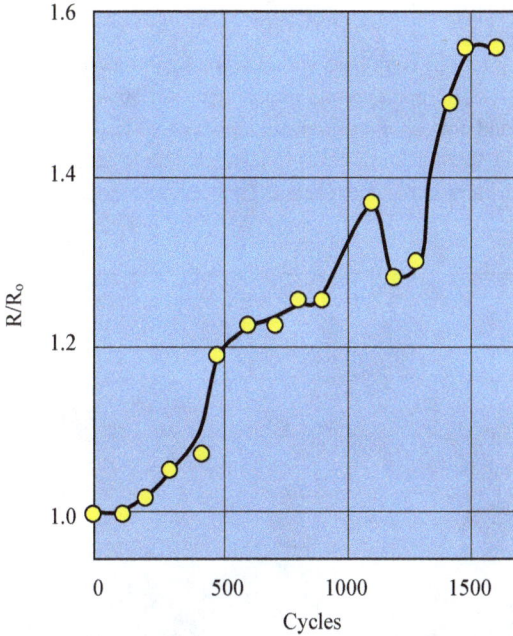

Figure 8. Sheet-resistance change of silver mesh, fabricated on a polyethylene naphthalate substrate, as a function of the number of bending cycles

A minimum resistivity of 7.13 x $10^{-8}\Omega$m was reached at an annealing temperature of 250C; 4.5 times greater than the bulk resistivity of silver. The discrepancy in resistivity was attributed to the presence of voids in the silver of the pattern. Comparable resistivities could be attained by using an annealing temperature of 200C and extending the annealing time to more than an hour. The lower temperature avoided any damage being caused to flexible substrates. The lowest sheet-resistance (1.79Ω/sq) was found for silver mesh with a line-width of 10Ωm. The mesh was very transparent in the visible light range, with transmittances ranging from 90.5 to 92% at a wavelength of 550nm, and a sheet-resistance of 16.4Ω/sq. The latter also resisted cyclic bending such that, after more than 1500 cycles with radii of curvature of between 1 and 20mm, the sheet resistance increased by less than 1.6 times. This was attributed to good adhesion between the silver nano-particles and the substrate. The line-width and pitch determine the transparency of the mesh, according to:

$$1 - (p-w)^2/p^2$$

The line-width (w) and pitch (p) were here varied at the same rate in order to maintain a constant fill-factor of 0.1; corresponding to a transmittance of 90%. Fill-factors generally range from 0.01 to 0.2 and correspond to transmittances of 80 to 99%. For a given line-width and pitch, a greater line-thickness leads to a smaller sheet-resistance, meaning that a higher thickness/width ratio is to be expected. The highest aspect ratio in the present case was 0.08.

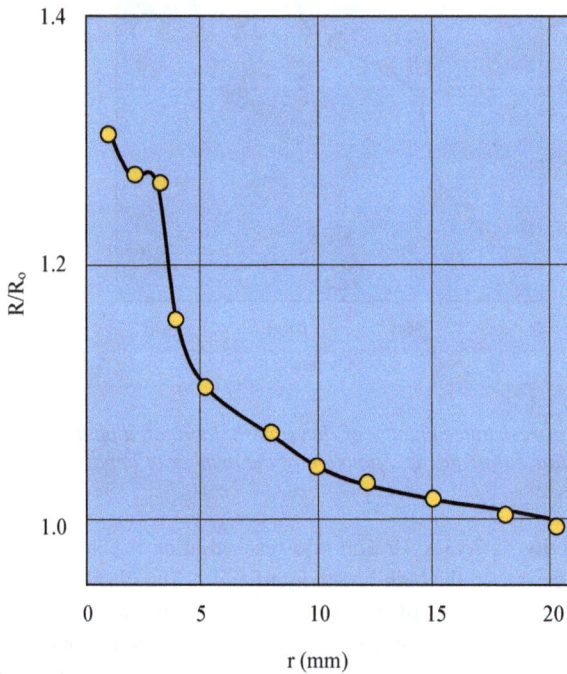

Figure 9. Sheet-resistance change of silver mesh, fabricated on a polyethylene naphthalate substrate, as a function of the radius of curvature, after 1500 cycles

Smart windows are those whose optical transparency can switch from highly transparent to opaque by incident solar illumination. Transparent conducting metal nano-mesh films are suitable candidates for the creation of thermochromic windows, due to their high thermal conductivity, high optical transparency at near-infrared wavelengths and great stability. Films of ZnO|Au|Al$_2$O$_3$ nano-mesh, with a periodicity of 370nm, exhibited[102] a transmittance of above 90% at 550nm and a sheet resistance of less than 20Ω/sq, thus out-performing competing nano-mesh films. The transparency of the smart window could be controlled by means of transient resistive heating so as to provoke thermochromic transition to the opaque state.

Nano-Material Preparation

One-dimensional metallic nano-structures such as nano-wires are of enormous utility with regard to electronic applications such as the present ones. The shape-anisotropy imparts novel optical capabilities to gold and silver nano-particles. These can include longitudinal plasmon-resonance bands in the visible and near-infrared regions of the spectrum. The methods used for the shape-controlled synthesis of silver and gold nano-crystals include chemical, electrochemical and physical methods. A widely-used method for the synthesis of nano-wires is one in which metal salts are reduced in an aqueous solution. This typically involves the use of a surfactant as a directing agent which can introduce asymmetry into the nano-crystal's shape. Variations in the concentration of the precursor salt and the surfactant, plus the nature of the surfactant and the nature and concentration of the reducing agents, as well as the presence of external salts and the pH-level of the reaction-solution all affect the nano-crystal's shape and size. The size and shape of the nano-crystals then affects the position of the plasmon bands.

The directed synthesis of single-phase nano-alloys is possible in systems of immiscible metals such as Rh-Pd, Rh-Pd-Pt, Au-Pt, Au-Ir, Au-Rh, Au-Ir-Rh and Ru-Pd, by using double complex salts as precursors[103]. Isostructural double complex salts can be used to prepare solid solutions which contain 3 or more metals. These can then be used to prepare 3- and multi-component nano-alloys with various compositions. Specially selected modes of thermal decomposition, at low heating-rates, permits the completion of metal reduction and nano-alloy formation below 250C. This prevents decomposition of the final metastable nano-alloy. During the thermolysis of multicomponent precursor compounds, nano-alloys form via various mechanisms, depending upon the electrochemical potential of the complexing metals and the nature of their ligand environment. The high thermal stability of metastable nano-alloys is attributed to the small size of the crystalline domains and to the stabilizing effect of graphene-like shells which were built up by layers of polymeric compounds and carbon.

The selective transformation of various starting materials by means of differing metal catalysts under optimized reaction conditions, leading to structurally different intermediates and products, is an essential approach to the generation of various molecular scaffolds. A given starting material can be exposed to a catalyst, leading to a common intermediate and differing scaffolds, by adjusting the reactivity of the metal catalyst. Porous cobalt assemblies, for example, can be prepared by using cobalt oxalate as a self-sacrificing precursor template, with *in situ* hydrogen reduction[104]. The resultant cobalt assemblies inherit the microstructure of the precursor and contain many pores due to gas-release. Such assemblies comprise hexagonal close-packed and face-centred cubic phases, and the proportions can be controlled by varying the reduction temperature. Hexagonal close-packed phases favour dielectric loss while the face-centred cubic phase results in marked magnetic losses. An optimum composition could combine both properties.

Copper micro-particles can be produced at room temperature by using a lyotropic liquid crystal template[105]. The latter has hexagonal ordering and is prepared by using a mixture of non-ionic surfactant and water in a weight ratio of 40:60. The controlled growth of copper particles is promoted in this medium by reducing cupric chloride using hydrazine hydrate under basic conditions. Monodisperse platelet-like copper micro-particles of about $0.25\mu m$ formed, well-dispersed in the lyotropic phase, with no aggregation. Such particles could remain stable for months in the liquid crystalline medium. Copper micro-platelets exhibited a great catalytic and electrocatalytic ability.

Using continuous electrodeposition, metallic nano-wires were produced which were smooth and uniform and had a face-centred cubic monocrystalline structure. Following interval-electrodeposition, the nano-wires were bamboo-like or pearl-necklace like and had a face-centred cubic structure. The length of nano-particle nano-wires or of monocrystalline nano-wires could be controlled by changing the cycle time or the continuous deposition time. Macroporous networks comprising interwoven carbon fibres which were loaded with Cu-Ni nano-particles of high ($538m^2/g$) surface area were prepared by template-directed synthesis with tissue paper acting as a bio-template. The effect of additions upon the dispersion and morphology of nickel-based and copper-based nano-dimensional products produced using chemical means was considered under conditions of directed synthesis[106]. Reduced nano-particles inherited the shape of the starting materials, with a corresponding increase in size. In going from hydroxides to oxides, the specific area increased. Gold nano-particles of controllable size and morphology could be produced[107] by using ordered mesophase templates that consisted of iso-octane, sodium bis(2-ethylhexyl) sulfosuccinate and lecithin; together with an aqueous phase which contained auric acid, $HAuCl_4$, as the gold precursor. Highlyfacetted nano-particles were formed upon directly reducing $HAuCl_4$ using the di-octyl sulfosuccinate termini of sodium bis(2-ethylhexyl) sulfosuccinate. The rapid reduction of $HAuCl_4$, by adding sodium borohydride to the aqueous phase, produced spherical nano-particles. Their size could be tailored by varying the auric acid concentration and the volume fraction of aqueous phase. Well-shaped octahedral nano-particles were prepared by using a one-pot non-seeded method. The most stable plane, {111}, could be perfectly formed in a template-directed environment which contained a very high concentration of cetyltrimethylammonium bromide surfactant, with ascorbic acid as the main reducer. Concave gold nano-crystals having various shapes and sizes have been prepared[108] by means of seeded growth. The process began with gold seeds having a well-defined morphology and a uniform size, but cubic and rod-like gold nano-crystals with notable concave features could then be derived. One-dimensional gold nano-structures have been prepared by using cobalt particles as sacrificial templates[109]. The synthesis of complex multicomponent colloidal nano-structures was made possible by using bi-functional polymers as geometry-directing agents[110] and stabilizing ligands. As well as gold, palladium and silver might be used as seeds to generate more complicated hybrid nano-structures when combined with cation exchange, the Kirkendall effect and oxidative etching. A one-step top-down approach was proposed[111] in which sinapinic acid-induced formation of nano-clusters involved a 3-step reaction process. Large (>200nm) gold nano-particles were first quickly formed after mixing sinapinic acid and an Au^{3+} precursor solution. Excess sinapinic acid molecules then self-assembled on the nano-particle surface, and large gold nano-particles were etched into small nano-particles by electrostatic repulsion between the neighbouring sinapinic acid molecules. Sinapinic acid -induced core etching of the nano-particles finally resulted in the formation of gold nano-clusters within 70min. The morphology of gold nano-structures strongly depended upon the molar ratio of 2-thiopheneacetic acid to $AuCl_4^-$. At lower ratios, uniform rosette-like micro-particles appeared which consisted of 30nm-thick gold nano-plates. These had a triangular prismatic or hexagonal geometry, with many defects. Upon increasing the above ratio, gold nano-particles or nano-rods which were densely surrounded by polythiophene polymers were produced. A controlled growth of flower-shaped gold crystals on rigid substrates was based upon a combination of soft nano-porous templates and multi-stage aqueous chemical methods[112]. An hexagonal array of gold nano-particles was first prepared by using a nano-porous thin membrane and a seed-mediated colloidal process. The size and morphology of the gold nano-flowers were further controlled by means of site-selective heterogeneous nucleation and growth onto the gold precursors. In a similar manner, a soft nano-porous template was used firstly to create an hexagonal array of silver nano-particles by using a seed-mediated growth colloidal process[113]. The size and morphology of the silver nano-crystals were then controlled by site-selective heterogeneous nucleation and growth onto the gold precursors. Removing the nano-porous polymer mask before growth led to the formation of flower-shaped silver crystals. Retaining the mask led to the formation of sheet-like structures. Control of the lateral dimensions of these metallic crystal arrays was possible by varying the seeding process and the temperature. The average interparticle distance of the ordered silver nano-crystals did not correspond exactly to that of the gold seed array, thus suggesting that the colloidal structures were not correlated. The evolution of silver ions into anisotropic crystals was driven mainly by the pre-formation of metal seeds on the substrate and by the polymer mask, which could kinetically control the growth of the nano-crystals.

Gold nano-tubes have been prepared via template-directed synthesis in porous alumina substrates by radio-frequency sputtering[114]. The resultant composites were then treated with acidic or alkaline aqueous solutions in order to remove the membrane, leaving self-supporting gold nano-tubes. This method permitted the preparation of composites and of free-standing metal nano-structures whose aspect-ratio could be controlled by varying the preparation conditions and the alumina-membrane pore-size. The possibility of preparing metal-oxide-metal heterojunction nano-wires in the $Au-TiO_2-Au$ system was tested by the sequential electroplating of gold and the electrodeposition of amorphous titanium oxide within the nano-holes of anodic aluminium oxide templates[115]. The template was then dissolved and the separated nano-wires were heat-treated so as to cause crystallization of amorphous TiO_x into nanocrystalline TiO_2. This nanocrystalline TiO_2 consisted mainly of the orthorhombic columbite phase. Closely-packed hierarchically-branched stable gold nano-wires were prepared[116] by using *Escherichia coli* cells and a seedless hexadecyltrimethylammonium bromide-directed method. The *Escherichia coli* cells played a double role in the biosorption of gold ions, and acted as preferential nucleation sites for gold nano-crystals during formation of the gold nano-wires. The correct hexadecyltrimethylammonium bromide concentration, plus a small excess of ascorbic acid, were essential for the formation of the nano-wires. Preferential nucleation sites, which were simultaneously mediated by adjacent cells, favoured branched growth. Random growth of a given nano-wire, having multiple branched points, produced hierarchically-branched nano-wires. The gold-ion adsorption was comparatively high for the gold-binding M13 spheroid, and this probably explained the differing morphologies.

Gold nano-rods have been prepared[117] by using a seed-mediated sequential growth technique that involved the use of citrate-stabilized seed crystals, which were then grown in solutions which contained $[AuCl_4]^-$, ascorbic acid and cetyltrimethylammonium bromide surfactant. The resultant nano-rods consisted of two superposed pairs of crystalline zones. These were either <112> and <100> or <110> and <111>. They were consistent with a cyclic penta-twinned crystal having five (111) twin boundaries which were arranged radially with respect to the [110] elongation direction. The nano-rods had an idealized 3-dimensional prismatic morphology, with ten (111) end-faces and five (100) or (110) side-faces … or possibly both. The seed crystals were initially transformed, by growth and aggregation, into decahedral penta-twinned crystals with 4% becoming elongated when a new reaction solution was added. The remaining twins grew isometrically. The repetition of this process increased the length of existing nano-rods, induced the further transformation of isometric particles (to produce second and third populations of shorter and wider nano-rods) and increased the size of the isometric crystals. A symmetry-breaking in face-centred cubic metallic structures, which produced anisotropic nano-particles, was based upon twinning that was greatly affected, during growth in solutions, by the adsorption of Au^I-surfactant complexes on the side-faces and edges of the isometric penta-twinned crystals. This was responsible for the preferential growth which occurred along the common [110] axis. It was suggested that the coupling of multiple-twinning, and habit-modification was a general mechanism that operated in many other processes for the preparation of metallic nano-particles having a high aspect-ratio.

Nano-cables with insulating silica shells on metallic gold nano-ribbons were prepared by means of peptide directed synthesis[118]. This was done by using the peptides, Midas-11 and Midas-11C to create gold nano-ribbons and nano-platelets, respectively, while Si#6-C was used to bind and coat silica onto the gold nano-structures. One thiol group in one cysteine of the C-terminal end of the Si#6-C peptide was sufficient to bind firmly to the gold nano-structures while not preventing the formation of a thin amorphous silica layer on the gold nano-structures.

Highly-branched gold-palladium ($Au_{46}Pd_{54}$) so-called nano-brambles were made by using a 1-pot aqueous method, with thymine being used as a weak stabilizing, capping and structure-directing agent[119]. These nano-brambles exhibited better surface-enhanced Raman scattering as compared with that of $Au_{30}Pd_{70}$ or $Au_{60}Pd_{40}$ nano-clusters. This was attributed to the unique structure and morphology of the $Au_{46}Pd_{54}$ nano-brambles, and to a synergism between gold and palladium.

Hydrogen bubbles have been used[120] as dynamic templates for the one-pot wet-chemistry preparation of large-scale self-supported bimetallic AuPt nano-wire networks with tunable compositions. The hydrogen bubbles were generated *in situ* by the hydrolysis and oxidation of sodium borohydride. The resultant AuPt nano-wires had clean surfaces because the gas bubbles did not require the use of acid/base or organic solvents for their removal. The as-prepared AuPt nano-wire networks were excellent electrocatalysts and were durable with respect to ethanol oxidation and oxygen reduction reactions.

It was possible to synthesize FePt nano-particle chains by refluxing platinum acetylacetonate Pt(acac)$_2$ and iron acetylacetonate Fe(acac)$_3$ in pentandiol in the presence of polyethelenimine cationic polyelectrolyte, with succinic acid as a surfactant. The latter two components controlled the size and morphology of the nanostructures, and the nano-wire diameter could be varied from 20 to 40nm via particle-size control while the lengths ranged from 200nm to several microns. Following annealing at 550C, the interconnected FePt nano-particles coalesced to form polycrystalline nano-wires of L1$_0$ FePt.

A two-dimensional strategy was proposed[121] for the template-directed synthesis of one-dimensional kink-rich Pd$_3$Pb nano-wires with numerous grain boundaries. Ultra-thin palladium nano-sheets were first produced so as to serve as self-sacrificial two-dimensional nano-templates. Dynamic equilibrium growth was then established on the two-dimensional palladium nano-sheets via the etching of palladium atoms and the edge-preferred co-deposition of Pd/Pb atoms. This was then followed by the oriented attachment of Pd/Pb alloy nano-grains. High yields of kink-rich Pd$_3$Pb nano-wires, having numerous grain-boundary defects were obtained.

Platinum nano-particles and nano-wires were produced [122] by using honey in a bio-directed synthesis method. The conversion of platinum ions into 2.2nm nano-particles was possible at 100C in aqueous honey solution. Longer heating times produced nano-wires which were 5 to 15nm in length and formed via the self-assembly of platinum nano-particles. The latter were highly crystalline and face-centred cubic. It was proposed that the platinum nano-particles were bound to protein due to the carboxylate ion group.

Platinum-copper nano-wires were prepared at room temperature by using a 1-step glucose-directed surfactant-free method in which CuCl$_2$ and H$_2$PtCl$_6$ were co-reduced by sodium borohydride[123]. The glucose acted as a 1-dimensional growth-directing agent and as a stabilizer. The average diameter of the nano-wires was about 3.1nm, and they had a face-centred cubic structure.

Nano-cubes and octahedra of PtNiFe were prepared by adjusting alloy compositions and controlling the effect of crystal-facet/surfactant binding upon growth seeds[124]. Nano-wires grew in cylindrical templates, built by using high concentrations of oleylamine.

Silver nano-wire arrays of regular and uniform size have been created[125] within the nano-channels of anodic aluminium oxide templated by using a paired-cell method. The as-synthesized samples were composed of face-centred cubic structures with an average diameter of 60 to 70nm. The nano-wires had a preferred monocrystalline structure. The

Materials Research Forum LLC

https://doi.org/10.21741/9781644903476

spectrum of the nano-wire arrays exhibits an ultra-violet emission band at 383nm which was attributed to a transverse dipole resonance of the arrays. A good surface-enhanced Raman scattering spectrum was observed upon excitation with a 514.5nm laser, and the intensity of the peak was some 23 times higher than that from an empty template.

Hierarchical silver assemblies have been synthesized in solution by using small acid molecules (citric, mandelic, etc.)[126]. This acid-directed self-assembly of metal nano-particles into large entities having complex structures could be achieved without requiring polymer surfactants or capping agents. The assembled structures had very rough surfaces, and core-shell silver wires exhibited a particularly high surface-enhanced Raman scattering sensitivity toward melamine, with no obvious polarization-dependent behaviour. Silver nano-wires were prepared by using a 1-pot method and transparent conductive electrodes were made by spin-coating[127]. Thin-film electrodes with a diameter of 80nm, length of 45μm and thickness of 78nm could have a sheet resistance as low as 83.2Ω/square, while the optical transmittance could be as high as 92.8% at 550nm. The figure-of-merit, ratio of transparency to sheet resistance, could thus be as high as 0.011. The transparent conductive electrodes which were prepared by using the silver nano-wires exhibited an excellent visible-wavelength transparency between 450 and 750nm, with an optimum transparency value of up to 97%.

The concatenation of silver nano-wire networks by ion-bombardment is a useful process for opto-electronics and nano-electronics. Silver nano-wires were bombarded with MeV copper ions and the effects of ion fluence upon ultra-violet and visible transmittance ranges and electrical properties of the copper-ion bombarded silver nano-wires were investigated[128]. Rietveld X-ray powder diffractograms of the non-bombarded and ion-bombarded materials revealed that the lattice constant and unit-cell volume were slightly changed by the bombardment; the lattice parameter had increased from 4.028 to 4.031Å. No impurities were detected that might affect the stability of the structure at high fluencies. Indeed, with increasing dose, the full-width at half-maximum of peaks decreased somewhat. This indicated that the material became more crystalline, with the crystallite size increasing from 64.5 to 71.4nm. This was attributed to bombardment-induced heating of the nano-wires. The electrical conductivity (table 12) and optical transmittance increased with increasing fluence. At 1×10^{15}ions/cm^2, the optical transmittance of silver nano-wire thin films increased to 34% in the visible range and to 19% in the ultra-violet range, as compared with non-bombarded silver nano-wire thin films. For a given dose, the electrical conductivity increased to twice its pristine value, and the increase in optical transmittance was attributed to ion beam induced localized heating, which caused slicing of the nano-wires while beam-induced fusion of the nano-wires at their contact points was the main reason for the increase in electrical conductivity.

Table 12. Direct-current conductivity of silver nano-wire networks as a function of ion dose

Dose (ions/cm^2)	Conductivity (mS/sq)
0	40
2.5×10^{12}	42
3.1×10^{13}	52
4.5×10^{13}	68
5.1×10^{14}	78
6.5×10^{14}	82
7.1×10^{15}	76

Silver nano-rods having various polydispersions were synthesized[129] in cetyltrimethylammonium bromide rod-shaped micelles by inducing the oriented growth of silver seeds and adjusting the cetyltrimethylammonium bromide content, leading to formation of the nano-rods within 600s. The optimum volume of 0.1M cetyltrimethylammonium bromide was 15.0m*l*, given that the volume of cetyltrimethylammonium bromide added was a key factor governing the dispersion of the nano-rods. The aging time was also important in controlling the morphology, due to the oxidation of the silver nano-rods by Br$^-$ and O_2 and due to their Ostwald ripening. Ablation of the tops of longer nano-rods was commonly associated with the growth of shorter nano-rods and of nano-spheres. The size distribution of the nano-rods could thus be more uniform in the early stages of aging. All of the nano-rods in colloidal solution were expected to turn into nearly spherical nano-particles of large size. Large-scale silver triangular nano-plates were synthesized[130] by reducing aqueous silver nitrate with sodium borohydride in the presence of sodium citrate and dioctyl sulfosuccinate sodium salt. The nano-plates were monocrystalline, and the optical in-plane dipole plasmon band of the nano-plates extended to about 1230nm (near-infrared). Varying the various reagent concentrations, pH-levels and reaction times showed that triangular nano-frame growth was possible only under particular experimental conditions. At relatively high concentrations of glycyl glycine, the molecules could act as both reducing agents and capping ligands when preparing silver nano-plates in solution[131]. The silver was initially obtained using a glycine/AgNO$_3$ ratio of 2:1 (glycyl glycine concentration of 2.7mM). The size of the nano-plates increased to 2μm as the molar ratio of glycyl glycine was reduced. Nano-particles alone were produced at a glycyl glycine concentration of 0.67mM. It was concluded that preferential adsorption of glycyl glycine on the {111} plane of the silver crystals played an important role in stabilizing that plane, leading to nano-plates having the {111} plane as an upper face. Nano-plates with triangular, hexagonal and truncated triangular shapes were also obtained when using peptides such as alanyl glycine as templates. When the preparation temperature was 130C, the main product was nano-particles. It was suggested that low temperatures could not provide enough energy to activate the faces required for the anisotropic growth of nano-plates. The yield of silver nano-plates relative to the total number of nano-particles could be as high as about 80%. The ratio of glycyl glycine to AgNO$_3$ was the key factor in forming silver nano-plates. A general procedure has been described, for the preparation of hybrid nano-particles, which permits the nanostructure morphology to be chosen simply by changing the metal anion while keeping the other conditions constant[132]. Both Ag/Cu$_2$ZnSnS$_4$ core-shell nano-particles and Ag$_2$S-Cu$_2$ZnSnS$_4$ Janus nano-particles, as well as PbS and Au/AuAg hybrid analogues, could be synthesized. Nucleation of the semiconductor was the critical step in the synthesis of a given hybrid. The formation of Ag/semiconductor core–shell nano-particles was affected by the presence of chloride ions in the solution. These improved the nucleation of the semiconductor phase on the surface of the metal core while subsequent growth of the Cu$_2$ZnSnS$_4$ resulted in the formation of a core–shell morphology. The resultant hybrid nanostructures had a non-epitaxial interface but were highly crystalline. The growth mechanism depended upon carefully balancing the reactivity of the metal precursors, the thiol precursors and the surface chemistry of the seeds. Changing the metal precursors from chlorides to acetates led to the formation of Ag$_2$S–semiconductor Janus nano-particles. The use of metal acetates led to the self-nucleation of Cu$_2$ZnSnS$_4$, which was a result of the lower reactivity of the acetate precursors when compared with chlorides. Silver was converted to Ag$_2$S in the presence of sulphur and was adsorbed on the Cu$_2$ZnSnS$_4$ surface to form an elongated Janus-like morphology.

Silver nanowire-based transparent micro-electrode arrays and interconnects have been designed to be high-performance flexible component which can be seamlessly integrated with soft tissues[133]. The nano-wire arrays possessed an optical transparency which was greater than 90.0% at 550nm, and were mechanically stable at up to 100000 bends of

5mm in radius. The micro-electrodes had a normalized electrochemical impedance of 3.4 to $15\Omega cm^2$ at 1kHz, and the interconnects had a sheet resistance of 4.1 to 25Ω/sq. These structures were also biocompatible.

High-resolution printed metal grids are an alternative to indium tin oxide as the transparent conductive electrode which is a key component of organic light-emitting diodes. A study was made[134] of evaporated diodes having a printed copper grid with a line-width of less than 3µm; the use of copper making the grid cheaper than one based upon gold or silver. One problem is that the protruding nature of metal grids can cause electrical short-circuits, and to a decrease in the emission intensity within the area enclosed by the metal grid lines. A highly-conductive charge-injection layer which is tens or hundreds of nanometres thick can planarize the metal grid and increase the uniformity of the spatial luminance. An inkjet-printed silver grid is too obtrusive, because of its line-width of more than 100µm. Another approach to ensuring uniform spatial luminance is to decrease the gap between the metal grid-lines. An organic light-emitting diode with a printed copper grid was prepared by using a thick evaporated doped hole-injection layer, and was compared with an equivalent device which was based upon indium tin oxide. The spatial luminance distribution was measured, and compared with finite-element simulations. The line-width, pitch and line-thickness of the printed copper grid ranged from 1.0 to 3.5µm, from 15 to 30µm and from 50 to 210nm, respectively. The sheet resistance and visible light transmittance ranged from 13 to 56Ω/sq and from 72 to 80%, respectively. The transmittance of the printed copper grid was flat within the visible-light region and was very close to that of indium tin oxide. The transmittance of the doped hole-injection layer, within the visible-light range, decreased monotonically with increasing layer thickness. A copper-grid diode without a doped hole-injection layer did not emit light, and suffered from electrical short-circuits. Copper-grid diodes with a greater than 50nm doped injection-layer exhibited uniform light emission. The luminance and efficacy were lower for copper-grid diodes than for comparable indium tin oxide diodes; particularly at low voltages. The hole-injection and electron-injection layers were the same for both diodes, and the difference was therefore ascribed to non-ohmic contact between the copper and the doped hole-injection layer. At higher voltages however the injection barrier was overcome and the current densities and luminances of both diodes were similar. Both diodes exhibited a good macroscopic uniformity. A difference in the overall current, voltage and light characteristics, when compared with indium tin oxide devices was attributed to the effect of shadowing by the metal grid and a limited barrier to injection from the copper grid and into the doped hole-injection layer. The luminance in the open areas between the copper grid-lines could be quite-well predicted, but the emission intensity at the grid itself was much higher expected. This was due to

Transparent Metals   Materials Research Forum LLC
Materials Research Foundations **174** (2025)   https://doi.org/10.21741/9781644903476

unconsidered optical cavity effects. The emission intensity close to the grid was also lower than expected, but this did not seem to be related to optical out-coupling or morphology.

Transparent indium tin oxide electrodes have limitations in flexibility and stretchability, so alternatives such as metal films, metal nano-wires and conductive meshes have appeared. Few of them can simultaneously offer high flexibility, stretchability and good opto-electronic properties. Nano-fibre, being a continuous and very long one-dimensional conductive material, is an ideal candidate for use as transparent electrodes due to its unique structure. Nano-fibres can be fabricated from metal/metallic compounds, carbon-based materials and conductive polymers, etc. The metal-based materials offer the advantages of simple preparation and low resistivity, but exhibit poor bending and stretching behaviours. Light-scattering also has a marked effect upon the photoelectric properties of the electrode. By adding other materials to nano-fibres, or by encasing them, it is possible to improve electrical conductivity, corrosion-resistance, bending and stretching. By depositing a layer of inert material onto the nano-fibre surface, the stability of electrodes can be greatly improved. Metals such as gold and silver can be incorporated into transparent electrodes and the latter's resistance decreases sharply at a critical volume ratio of the conductive materials. This ratio is governed by percolation theory, and depends largely upon the aspect ratio of the conductive material. By using conductive materials having a higher aspect ratio, a conductive path can be created throughout the entire specimen at a lower critical volume-ratio. So in order to create electrodes having both good conductivity and transmittance it is necessary to use metal nano-materials which possess a high aspect ratio. One-dimensional silver nanostructures possess a high conductivity (6.3×10^7S/m), together with moderate flexibility and high transmittance. The effectively infinite length of silver nano-fibres can be made into large-area 2-dimensional and 3-dimensional flexible electrodes. Cheaper copper nano-fibres can offer a transmittance of 90% and a sheet-resistance of 50Ω/sq. They are also very stretchable and flexible. On the other hand, copper has a relatively poor stability and is easily oxidized. Stretching is the most problematic deformation mode, and can be sub-divided into fibre straightening, fibre stretching, fibre reorientation into the tensile direction and inter-fibre sliding[135]. There are then 3 principal factors which are responsible for the increase in length during stretching. At first there is straightening as the radius of curvature of the bent fibre gradually increases under the tensile load. Then there is lengthening of the fibre itself. Thirdly there is a fibre-reorientation along the axis of stretching which reduces the angle between the preferred fibre orientation and the stretching axis. As this occurs, a shear strain is created between the nano-fibres. At a critical shear deformation, the shear stress at the interface exceeds the cohesive strength of the interface between the nano-fibres and the latter slide against one another. Metal nano-fibres possess good opto-electronic properties but poor tensile properties, so that some form of support is required. The properties of the various electrode materials have been summarized[136] (table 9).

Table 13. Properties of various transparent metal electrode materials

Material	Sheet Resistance(Ω/sq)	Transparency(%)	R
Ag	7	96	1306
Ag	3.63	97.57	4196
Ag	1	90	3485
Cu	2	90	1742
Cu	10	95	726
Cu	17	97	723
Cu	<0.9	>84	2299
Cu	0.058	90	60082
Pt	8.3	80	192
Au	10	91	390
W	0.2	90	17424
Ag/C	9.6	80	166
TiN/Cu	34	82	53
PAN(core)/Au(shell)	25	81	68
Cu(shell)/PAN(core)	0.42	97	29246
Au(core)/MnO$_2$(shell)	9.58	86	251
Cu/Zr	3.8	90	917
PDMS/Cu	8	92	553
PEDOT:PSS/Ag	2.12	84.65	1023

Au/C	2.7	91	1446
Ni:Co-S/Cu@Ni	12.19	89	258
Ag/PVB	10	90	348
Cu/P4VP	15.6	82	116
NiO$_{NP}$/Ag	16	75	76
PEDOT:PSS/MoO$_{3,NW}$/Ag$_{NF}$	9.7	82.8	196
PVA/Ag	0.9	84	2299
PVA/Ag	1.68-11.1	>70	575
PVA/Ag	2.56	90	1361
Ag$_{NW}$/polyamide	8.2	84.9	270

R: ratio of electrical conductance to optical conductance, PAN: polyacrylonitrile, PDMS: polydimethylsiloxane, PEDOT: poly(3,4-ethylene dioxythiophene), PSS: polystyrene sulfonate, PVB: polyvinyl butyral, PVA: polyvinyl acetate, NP: nano-particles, NW: nano-wires, NF: nano-fibres

Silver nano-wire networks are very good transparent electrodes for flexible optoelectronics, but suffer from instability of the nano-wire junctions and from high surface roughness. In one process[137], the entire substrate is exposed to silver vapour via thermal evaporation, so that the metal is selectively deposited onto a nano-fibre network. The contact resistance between the nano-wires is then zero and the surface roughness is suitably low. Nano-network electrodes can be prepared in a bottom-up manner in which an electrospun polymer nano-fibre network acts as a template for electroless plating or vacuum deposition of metal, followed by selectively transfer of the metallized polymer network to another substrate. They can alternatively be prepared by using a top-down method in which a polymer nano-fibre network is spun onto a thin metal film so as to act as a metal etching mask. Upon etching away the metal which is not covered with polymer nano-fibres, the latter are removed so as to leave a network of metal lines of uniform height and zero junction-resistance. Junction-free networks are preferred because the contacts are prone to oxidation and localized heating. Only nano-network electrodes which are prepared by using the top-down method have generally proved to be suitable as the substrate electrode. Those which are prepared by using the bottom-up approach have a too-high surface roughness. A different scalable approach to the preparation of junction-free transparent silver nanowire-network electrodes avoids any complex metal nano-network transfer or metal etching step. This approach exploits the fact that highly-fluorinated organic compounds can have an extremely low condensation coefficient for silver vapour. Transparent silver electrodes having an average transparency of 90.8% within the wavelength range of 400 to 800nm, and a sheet resistance of 6.3Ω/sq, can be prepared by using this method, in which polyvinylpyrrolidone nano-wires, doped with 3-mercaptopropyl trimethoxysilane and 3-aminopropyl trimethoxysilane, are directly electrospun onto flexible plastic substrates which are coated with a thin film of organofluorine polymer and small-molecule mixture: poly (3,3,4,4,5,5,6,6,7,7,8,8,9,9,10,10,10-heptadecafluorodecyl methacrylate) trichloro (1H,1H,2H,2H perfluorooctyl) silane. The diameter of the polyvinylpyrrolidone nano-wires is 294nm before metallization. The nano-wires then undergo fusion which reduces the surface roughness before the entire substrate is exposed to metal vapour so that a 100nm silver film selectively condenses onto the doped polymer wires so as to form junction-free metal nano-networks. Organofluorines generally have a very low surface tension, and solutions of small-molecule organofluorines have a very low viscosity. This makes it difficult to form a uniform thin film having a thickness of more than 10nm. Suitable blends having a very low condensation coefficient for silver can however be easily cast into thin films of controllable thickness on various substrates, and have a thickness of 130nm. All of the silver nano-wire networks which were prepared by using the method had a direct-current conductivity to optical conductivity ratio of between 600 and 800. This met the requirement of a 85% transmittance at a 10 to 15Ω/sq sheet-resistance, for photovoltaic applications. For a total transparency of 90.8%, the electrode sheet resistance is 6.3Ω/sq. This is comparable to the best reported value for metal nano-wire electrodes which are prepared by using the usual top-down etching methods. Cracks formed in indium tin oxide films after a few bending cycles, while the silver nano-wire films were stable even when a sharper film-bending radius was used. The silver nano-wire heated up at a lower applied voltage than did indium tin oxide; well below the typical operating voltage of 12V. The temperature of the nano-wire network also stabilized more quickly than did indium tin oxide, and produced a uniform heat distribution at a 5V bias; even upon bending. The sub-micron diameter of the nano-networks permitted a high degree of freedom in controlling the nano-wire density and the uniformity of the heat distribution. The photo-active layer thickness needs however to be increased to more than 300nm because of the high defect-density which is associated with much thinner bulk heterojunction layers, and because of the difficulty in producing a uniform layer thickness over large areas. Transparent conducting electrodes which are based upon metal meshes are the best candidates because of their inherently high conductivity, optical transparency and mechanical robustness[138]. A very stretchable transparent electrical heater was constructed[139] by partially embedding a silver-nanowire

Transparent Metals   Materials Research Forum LLC
Materials Research Foundations **174** (2025)   https://doi.org/10.21741/9781644903476

percolative network in an elastic substrate. The stretchable network could survive straining, bending and twisting. It is also useful to have transparent conducting materials which can block electromagnetic radiation, and to increase electromagnetic shielding without losing too much optical transparency. A composite electrode was created[140] by depositing a continuous and homogeneous ultra-thin 7nm gold film onto monolayer graphene, leading to an average sheet-resistance of 24.6Ω/sq and a transmittance of 74.6% at 520 nm. The excellent properties of the composite electrode were attributed to suppressed Volmer-Weber growth of the ultra-thin gold film by the hexagonal carbon lattice of the graphene. A rigid supercapacitor which was based upon the composite electrode had a capacitance of 81.5μF/cm^2 at a scan-rate of 0.1V/s. This was 17 times higher than that of a device which was based upon monolayer graphene. A flexible supercapacitor which was based upon the composite electrode could be bent to a radius of curvature of 2mm, with a slight decrease in performance after 1800 bending cycles. Highly-flexible and ultra-thin transparent conductive electrodes were created[141] for use in organic light-emitting devices by combining single-layer graphene with thin silver layers which were a few nanometres thickness. The as-prepared graphene|Ag(8nm) composite electrodes had a sheet-resistance of 8.5Ω/sq, exhibited high stability during 500 bending cycles and a 74% transmittance at 550nm. When the composite electrodes were used as anodes in organic light-emitting devices, they led to turn-on voltages of 2.4V, with a luminance greater than 1300cd/m^2 at 5V and a maximum luminance which attained 40000cd/m^2 at 9V. The devices also functioned normally with a less-than 1cm bend radius. Suitable flexible hybrid films can be constructed[142] by nano-printing metal meshes and adding a graphene coating. The balance between transmittance and shielding effectiveness can be adjusted by controlling the width and period of the metal mesh. A coating further aids this. When compared with plain metal mesh, the electromagnetic shielding effectiveness can be increased by sacrificing little optical transparency. As graphene can affect the homogeneity of the metal mesh at higher frequencies, a hybrid shielding film which involves graphene and metal mesh can function at microwave and millimetre frequencies. Shielding can be achieved by enhancing reflection or absorption, and conventional microwave absorbers and reflective coatings are opaque at visible frequencies. This is due to the properties of metals and lossy mixtures. Such shielding materials cannot therefore be used in observation windows. In order to produce transparent electromagnetic shields, optically transparent conductive films such as indium tin oxide have been used. Unlike absorption-type transparent electromagnetic shielding materials, homogeneous transparent conductive thin films without periodic patterns can provide reflection-type electromagnetic shielding. Increased microwave absorption is exhibited by randomly dispersed systems which contain conductive particles. Reflection-type transparent electromagnetic shielding can operate over a broader bandwidth than can absorption-type shielding because homogeneous transparent conductive films are insensitive to wavelength if their dimensions are suitably chosen. In order to balance optical transparency and microwave shielding, the thickness of semiconducting transparent films tends to be limited to the nanometre-scale. The shielding effectiveness of reflection-type transparent electromagnetic shielding materials is closely related to their sheet resistance, so metal-meshes and metal nano-wire networks are attractive for shielding.

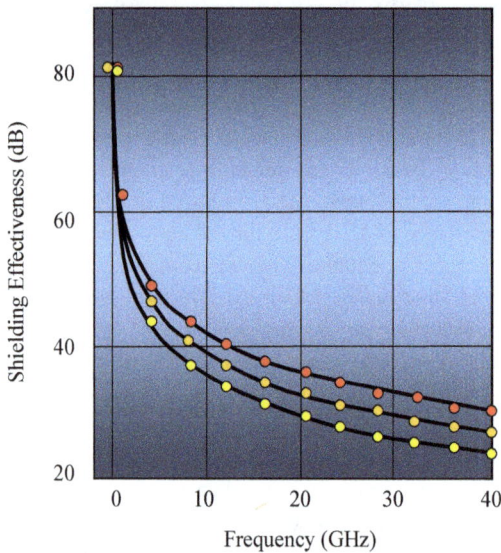

Figure 10. Shielding effectiveness of metal meshes with an optical transmittance of 85%. Red: square, orange: circular, yellow: hexagonal

The period of the metal mesh is usually far smaller than the microwave wavelength, and should be much greater than visible to infrared wavelengths in the visible region (figures 10 and 11). For a given periodicity and line-width, a metal mesh can offer a lower sheet resistance and better shielding if the height of the metal mesh is increased. Because the quality of microwave shielding is closely related to the conductivity of the material, graphene-coated metal mesh can provide high electromagnetic shielding. Mesh|graphene hybrid structures can be prepared by means of nano-imprinting and chemical vapour deposition. A hybrid structure can offer 9.5 to 11.5GHz with an average optical transmittance of 77.3% between 400 and 780nm. Simulations involved copper with a conductivity of 5.96 x 10^7S/m and 4 types of periodic structure: square, circular, hexagonal and triangular.

The transmittance of the 4 types of metal mesh was fixed at 85% to order to isolate the effect of shape. The period and thickness were 99.46 and 3.88μm, respectively, for the circular lattice, 64 and 5μm for the square lattice, 96 and 7.49μm for the hexagonal lattice and 154.7 and 6.07μm for the triangular lattice. The thickness of the metal mesh was always 5μm. A mesh with square cells was shown to offer a higher shielding effectiveness than did the other shapes. The effectiveness decreased when the frequency increased. The shielding effect of the square mesh improved when the thickness and line width were increased. The shielding decreased when the period increased. An increased line width or decreased period made the overall sheet resistance of the metal mesh smaller and thus led to increased shielding.

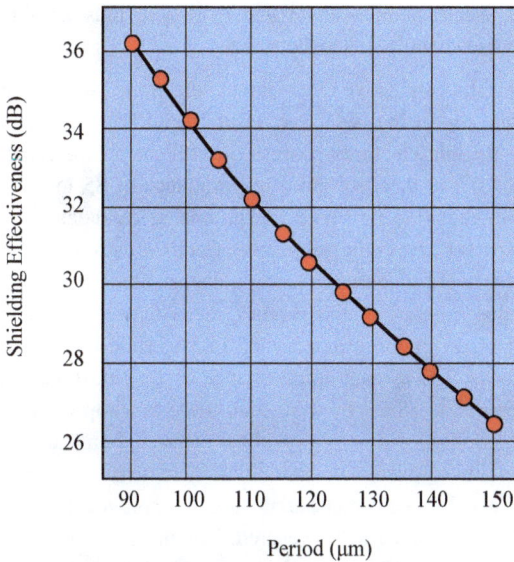

Figure 11. Shielding effectiveness of square-patterned metal mesh as a function of the period

The optical transmittance depends upon the period and line width. The period and line width of the square mesh varied from 100 to 120µm and from 3.5 to 5µm, respectively, and the optical transmittance increased when the lines decreased in width or the period increased. Having optimized the mesh, the next step is to add the graphene. The latter is prepared by using chemical vapour deposition and is transferred to the mesh film by using a temporary polymethyl methacrylate intermediate film. The optical transmittance of the hybrid films was measured for wavelengths ranging from 400 to 780nm. As compared with plain metal mesh, samples which were coated with graphene had an optical transmittance which was reduced by about 6%. The sheet-resistance of the reference sample was 0.43Ω/sq while those of the coated samples ranged from 0.26 to 0.40Ω/sq. The shielding effectiveness of the hybrid samples ranged from 26.2 to 31.8dB. The shielding results for metal mesh and hybrid films exhibit a similar trend to the simulated ones. A comprehensive study was made[143] of the electrical, optical and mechanical properties of hybrid nanostructures which were based upon 2-dimensional graphene and networks of 1-dimensional metal nano-wires. They offered a sheet-resistance of 33Ω/sq and a transmittance of 94% in the visible range, together with stability with respect to electrical breakdown and oxidation, plus 27% flexibility in bending strain and 100% stretchability in tensile strain.

An embedded nickel-mesh conductive transparent electrode was prepared[144] on polyethylene terephthalate by using selective electrodeposition and inverted film-processing. The durable flexible electrode possessed excellent photoelectric properties, with a sheet-resistance of 0.2 to 0.5Ω/sq and a transmittance of 85 to 87%. Ultra-thin silver films with a low percolation threshold are promising candidates as transparent electrodes for semi-transparent perovskite solar cells. Continuous films with percolation thresholds lower than 6nm were prepared[145] using a copper seed layer which imparted excellent conductivity plus broadband transparency. Semi-transparent cells with 6nm silver electrodes offered a conversion efficiency of 14.5% while retaining 88% of the performance of opaque cells. The average transmittance of the semi-transparent cells, at 800 to 1200nm, was improved to 58.6% by using zinc oxide as an optical coupling layer. The use of 5nm silver electrodes, combined with silver grids, and optical coupling layers with a higher refractive index, were anticipated to increase further the conversion efficiency of tandem solar cells. Semi-transparent flexible indium-free perovskite solar cells with a high visible transmittance were prepared[146] by using two types of composite ultra-thin metallic electrodes: MoO_3 | $AuMoO_3$ | Au and MoO_3 | Au | Ag | MoO_3 | Alq_3MoO_3 | Au | Ag | MoO_3 | Alq_3 as the bottom and top electrodes, respectively. These electrodes offered high electrical conductivity, mechanical robustness and a high optical transparency. An overall power conversion efficiency of 6.96% and an average visible transmittance of 18.16% at 380 to 790nm was possible. The devices retained 71% of their initial conversion efficiency after 1000 bending cycles with a bend radius of 4mm.

Ultraviolet ozone treatment[147] of molybdenum trioxide was a simple and efficient method for obtaining continuous and smooth silver thin films in thermally evaporated $MoO_3|Ag|MoO_3$ multi-layered structures for use as transparent electrodes for small-molecule organic solar cells. The ozone treatment could modify the non-stoichiometric MoO_3 surfaces so as to increase the Mo^{6+}/Mo^{5+} composition ratio and work function of MoO_{3-x}. The ozone treatment of the MoO_3 bottom layer improved the wettability of the silver on MoO_3 and enhanced lateral growth of the silver thin film; thus reducing the percolation threshold thickness of the continuous silver layer. Due to the formation of an ultra-thin silver interlayer with a continuous and smooth surface morphology, the multi-layered electrode after 180s of ozone treatment of MoO_3 offered a maximum transmittance of 89.1% and a sheet-resistance of $8.0\Omega/sq$. When an optimum ozone-treated $MoO_3|Ag(7.5nm)|MoO_3$ film was used as the anode in organic solar cells with a copper-phthalocyanine|fullerene planar heterojunction structure, the cells had a power conversion efficiency of 0.55%. This was 2.0 and 1.2 times higher than that of devices with untreated $MoO_3|Ag(7.5nm)|MoO_3(0.27\%)$ and $MoO_3|Ag(10nm)|MoO_3$ electrodes (0.46%), respectively.

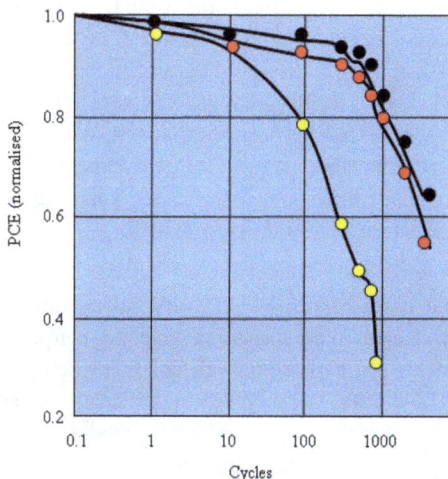

Figure 12. Power conversion efficiency of flexible perovskite solar cells, with various electrodes, as a function of the number of bending cycles at a radius of 4mm. Yellow: PET\ITO, red: PET\Ag-mesh, black: PET\Ni-mesh

Hybrid electrodes were used for perovskite solar cells which had good electrical properties together with notable environmental and mechanical stabilities. They retained 76% of their efficiency following 2000 bending cycles. Nickel was first electrodeposited into randomly distributed micro-grooves, in a photo-resist layer on an indium tin oxide substrate, generated by laser-lithography. A free-standing metal mesh thus remained on the oxide glass after removing the photo-resist. A coplanar film-inversion process then produced ultra-flexible embedded PET|Ni-mesh electrodes. The dense nickel mesh markedly improved the opto-electrical and mechanical properties of the transparent electrode. Detachment and migration of the metal particles were also greatly reduced during bending.

The power-conversion efficiency of flexible perovskite solar cells could attain 17.3%. FET|Ni-mesh solar cells retained 76% of their efficiency after 2000 bending cycles. Silver-mesh transparent electrodes were also prepared, showing that mesh made from nickel was relatively line-dense whereas silver lines formed via the accumulation of small loose silver particles, thus implying that the nickel mesh was more stable and mechanically robust. The best flexible device which was based upon a PET|Ni-mesh substrate offered the above power conversion efficiency of 17.3%, with a short-circuit current density of $21.78 mA/cm^2$, an open-circuit voltage of 1.02V and a fill-factor of 0.78. The PET|Ni-mesh devices out-performed PET|Ag-mesh-based devices. The topography of the nickel wire suggested that the area of perovskite above the nickel wire was more conductive and soon collected charges. The morphologies of the perovskite crystals above the nickel wires and the surroundings were not different but, due to stress in the substrate, a crack formed between the regions. When the perovskite film was exposed to air for a week, the morphology of the perovskite degraded out from the nickel wires and extended toward the centre of the mesh. This suggested that the substrate affected the environmental stability of the perovskite film.

Perovskite films having differing substrates were exposed to air for 3, 72 or 168h. For new films the morphologies of perovskite layers having differing substrates did not exhibit any apparent change. With increasing time the perovskite above silver wires exhibited decomposition when exposed for 72h, and this spread to the surroundings. The perovskite of the PET|Ni-mesh was relatively more stable. Silver particles in PET|Ag mesh electrodes could easily fall off and combine with iodine in the perovskite film so as to form AgI. This then accelerated decomposition of the perovskite. Because a dense oxide film was easily formed on the surface of the nickel mesh, that could hinder any further decomposition of the mesh and improve the stability of the device. The degradation of device efficiency was consistent with changes in morphology and, with increasing time, the efficiency of the nickel-mesh solar cell remained at the initial 86%

after 192h in storage. The silver-mesh cell efficiency meanwhile decreased to the initial 59%. Flexible perovskite solar cells which were based upon various substrates were bent to 5 different radii of curvature (∞, 8, 6, 4, 2mm). As the degree of bending was increased the performance of the device decreased (figure 12). The degradation of a PET|ITO device was greatest, whereas PET|Ni-mesh devices were relatively stable. Even at a bending radius of 2mm the efficiency of the cell retained 97% of its original value. Devices which were based upon PET|Ag-mesh and PET|ITO substrates retained only 95% and 87%, respectively. The power conversion efficiencies of perovskite solar cells on various flexible electrodes were compared and 5000 continuous bending cycles were performed to a radius of 4mm. Flexible cells which were based upon PET|Ni-mesh electrodes exhibited a promising bending stability in that, after 500 cycles, the efficiency remained above 92% of the original value. A PET|ITO-based device retained only 78% of the original value when bent 100 times. After 2000 cycles the PET|Ni-mesh device retained 76% of its initial efficiency. Changes in the structure of the perovskite on differing flexible substrates at various curvatures were also determined. In the case of PET|Ag-mesh perovskite crystals, the peak-width and peak-position of (110) X-ray diffraction signals changed significantly. Meanwhile the PET|Ni-mesh perovskite crystals were relatively stable. The perovskite in the PET|Ni-mesh device was very stable, while the perovskite film in the PET|Ag-mesh device exhibited precipitation of PbI_2 at various depths. This suggested that loose silver particles can affect the stability of the bulk of the perovskite.

A transparent metal-mesh heater was created[148] via the intense pulsed-light sintering of copper-based particle ink on a polyethylene terephthalate substrate. The mesh electrode, with a line-width of 9μm was produced by filling intaglio patterns with silver-coated copper-particle ink. The latter was then pulsed-light (7.5J/cm^2) sintered for 20ms, yielding an electrode of 85% transparency and sheet resistance of just 2Ω/sq. The 9V flexible heater could operate at up to 130C. In the case of pure-copper nano-particles, the resistance increased with time and this could amount to 45 and 400 times the initial resistance at 100 and 150C, respectively. In the case of silver-coated copper particles (table 10, figures 13 to 16), hardly any increase in resistance occurred. In order to create a transparent heater on a polyethylene terephthalate substrate, by using silver-coated copper nano-particles, sintering had to be possible at temperatures lower than the glass transition temperature of the polyethylene terephthalate. In the case of pure copper nano-particles, sintering barely occurred even when the energy was increased without increasing the substrate temperature, and there was no improvement in the electrical properties. This was so, regardless of the size of the copper nano-particles. In the case of silver-coated copper, or pure silver, particles the resistance decreased as the energy was increased.

Table 14. Sheet resistance of pulsed-light sintered metal-particle films

Particles	Energy-Density(J/cm^2)	Temperature (C)	Sheet Resistance(Ω/sq)
Ag-coated Cu	5.5	-	20
Ag-coated Cu	6.5	-	18
Ag-coated Cu	7.5	-	17
Ag-coated Cu	7.5	80	8.0
Ag-coated Cu	7.5	100	2.4
Ag-coated Cu	7.5	120	1.5
Ag	5.5	-	3.4
Ag	6.5	-	3.3
Ag	7.5	-	3.1
Ag	7.5	80	3.0
Ag	7.5	100	2.3
Ag	7.5	120	1.6

The initial resistances of silver-coated copper and pure silver particles were 20 and 3.4Ω/sq, respectively. When the sintering energy was increased to 7.5J/cm^2, the resistances fell to 17 and 3.1Ω/sq, respectively. If the substrate temperature was increased to 120C, for a given optical energy, adequate sintering was possible in the case of silver-coated copper particles and the resistance was decreased. Increasing the substrate temperature to 150C damaged the polyethylene terephthalate substrate. When the light was of sufficient intensity, oxidized pure copper particles underwent reduction due to additives in the ink and permitted sintering. During sintering using an energy of 7.5J/cm^2, cracks initiated in the copper lines when the substrate temperature was 80C. At a substrate temperature of 100C, the cracks extended further and damage to the substrate occurred 120C. When the pure copper particles were 500 to 800nm in size, cracking was somewhat delayed and occurred at a substrate temperature of 100C, and cracks propagated into the substrate at 120C.

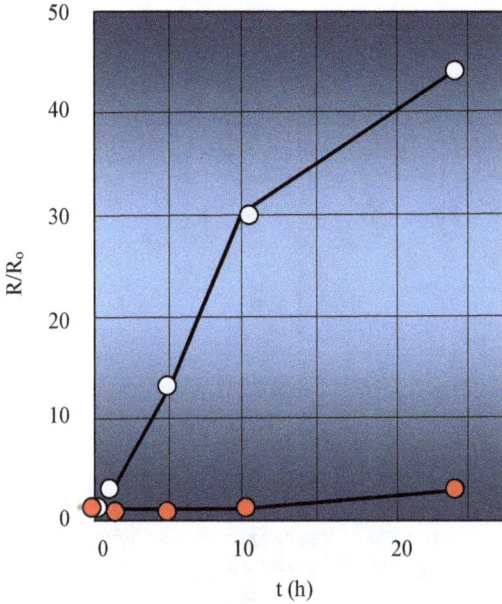

Figure 13. Resistance-increase ratio of sintered particles as a function of aging time at 100C. White: sintered pure copper particles, red: sintered silver-coated copper particles

When the particle size was smaller, the generation of reduction-heat increased due to the increase in the surface area over which sintering occurred and cracks therefore occurred at a lower substrate temperature, due to the accumulated heat. In the case of silver-coated copper particles, the resistance continued to decrease as sintering proceeded but no damage to the substrate or metal lines occurred, possibly because the silver coating on the copper surface prevented the latter's oxidation, thus reducing the reduction-heat of the oxide. In the case of pure copper particles, optical absorption occurred at wavelengths of 300 to 850nm. In the case of pure silver, or silver-coated copper, particles the absorption was much lower at wavelengths greater than 350nm. The lamp used for the sintering emitted light in the 350 to 850nm range. Pure silver or silver-coated copper particles thus absorbed less light and did not generate reduction-heat.

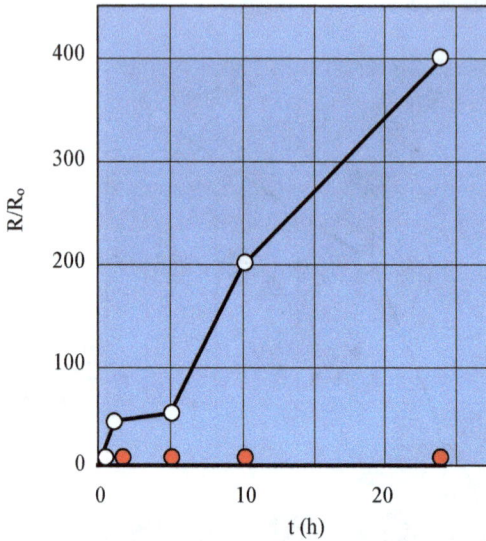

Figure 14. Resistance-increase ratio of sintered particles as a function of aging time at 150C. White: sintered pure copper particles, red: sintered silver-coated copper particles

The optical transmittance of silver-coated copper mesh was 86% at a wavelength of 550nm. The sheet resistances of silver-coated copper mesh, indium tin oxide and silver nano-wire film heaters were 2, 11 and 10.2Ω/sq, respectively. In the case of indium tin oxide and silver-nano-wire heaters, the currents as a function of temperature were similar, due to their similar resistances. In the case of a silver-coated copper mesh heater, the current change with voltage was much larger. The temperature of the metal-mesh heater attained 130C at a voltage of 9V while the indium tin oxide and silver-nanowire heaters reached only 70C. An easy method[149] for the fabrication of a metallic-grid transparent conductor on a flexible substrate was the selective laser-sintering of metal-nanoparticle ink. Metallic-grid transparent conductors with a transmittance greater than 85% and a sheet-resistance of 30Ω/sq were easily produced on glass and polymer substrates without requiring vacuum or high temperatures. Being a mask-less direct-writing method, the shape and other details of the grid could be easily changed. These metallic grids offered better stability with regard to adhesion and bending.

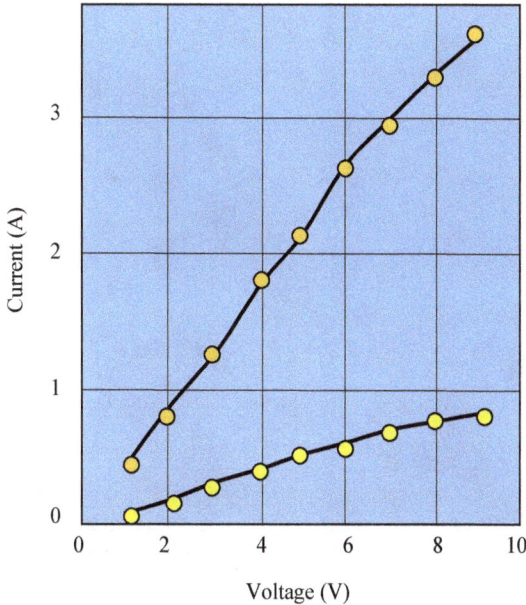

Figure 15. I-V characteristics of transparent heaters during operation. Orange: silver-coated copper mesh, yellow: silver nano-wire

Silver-nanowire stretchable transparent micro-electrodes (figure 17) have been produced[150] by using a simple selective-patterning method which permits the direct integration of nano-wire networks into elastomeric substrates. The resultant interfaces have a sheet-resistance of 1.52 to 4.35Ω/sq and a normalized electrochemical impedance of 3.78 to $6.04\Omega cm^2$, together with an optical transparency of 61.3 to 80.5% at 550nm and a stretchability of 40%. The silver nano-wire structures are sandwiched between a polydimethylsiloxane substrate and an encapsulation layer. The nano-wire network offers high transparency because of the open areas between the interconnected nano-wires.

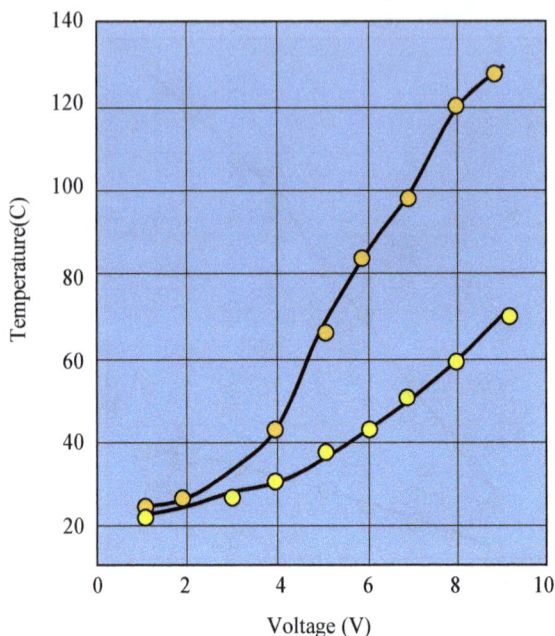

Figure 16. Temperature increase of transparent heaters during operation. Orange: silver-coated copper mesh, yellow: silver nano-wire

The stretchability meanwhile results from sliding between the nano-wires. The nano-wires are biocompatible and offer a much higher surface roughness than do planar metal micro-electrodes, thus increasing the effective surface area for electrophysiological monitoring. The transmittance of the silver nano-wire networks at 550nm increased from 61.3 to 75.2 and to 80.5% as the nano-wire concentration decreased from 20 to 10 and then to 8mg/ml. The pitch between micro-electrode sites was 2.15mm, and the dimensions of a single micro-electrode were 650µm x 650µm. When the nano-wires had a diameter of about 120nm and a length of between 10 and 20µm, the average sheet-resistance increased from 1.52Ω/sq to 3.77Ω/sq and 4.35Ω/sq, with the nano-wire concentration decreasing from 20 to 10 and to 8mg/ml due to the reduced silver nano-wire density in the conductive networks. Straining tests of nanowire/polydimethylsiloxane films were performed on 2.5cm x 3.5cm samples.

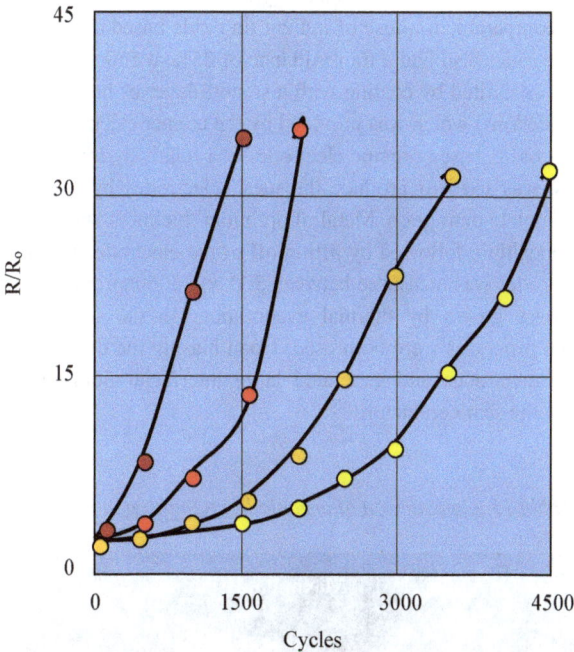

*Figure 17. Resistance change due to cyclic stretching of silver nano-wire electrodes.
Brown: 40% strain, red: 20% strain, orange: 10% strain, yellow 5% strain*

Following one-time stretching, the films had an unchanged resistance in the released state after up to 40% strain. The resistance remained stable for over 500 cycles of stretching to up to 20% strain. At 40% strain, the released films had a markedly increased resistance following 500 cycles of stretching and this was attributed to delamination of the nano-wire junctions under continuous high mechanical stresses. There was a clear decrease in the nano-wire network density, as compared with that of the pristine film. The resistance of the system remained unchanged after stretching, thus confirming the stable electrical behaviour of the networks following 500 stretching cycles at 10% strain.

In organic light-emitting diodes the transparent anode is typically made from transparent conductive oxides, but these have problems. Efficient indium-free transparent light-emitting diodes were made[151] by using gold, silver or copper mesh-based top electrodes. Copper electrodes with a sheet-resistance of about 7Ω/sq and uniform emission characteristics were created which had a relatively high current efficiency and luminance values which were comparable to those of indium tin oxide based devices (tables 11 and 12). The copper electrodes also had a transmittance of 85% at 550nm. Roughness due to the wire thickness was limited by coating with a smooth layer of hole-injection material. The luminance (4300cd/m^2) which was provided by the copper electrode was comparable to that of indium tin oxide based organic electrodes at a relatively lower turn-on voltage. Crackle-lithography was used to produce the mesh electrodes by first spin-coating a crackle precursor template onto glass. Metal of optimum thickness was then deposited by means of vapour deposition, followed by lifting-off of the electrode. The transmission of the transparent electrodes was measured between 300 and 1000nm. In one configuration, individual layers were grown by thermal evaporation. In the second configuration, solution and thermal processes were both used. Upon biasing the device, charge-carriers drifted towards the electrodes and generated large interfacial fields which promoted charge-injection into the semiconductor.

Table 15. Comparison of organic light-emitting diodes

Anode	Turn-On Voltage(V)	Maximum Brightness(cd/m^2)	Efficiency
ITO	4.64	4298	5.70
Ag-TCE	4.16	521	0.20
Au-TCE	6.16	103	0.71
Cu-TCE	5.61	796	0.99

TCE: transparent conductive electrode, ITO: indium tin oxide

Recombination which involved charge-carriers that were injected from the counter-electrode occurred only when near to the grid-lines. Metal meshes (50nm) having optimum values of transmittance and resistivity of 84% to 87% and 7 to 20Ω/sq were created. The relative spectral intensity of copper transparent electrode devices was much greater than that of silver or gold based devices. The better performance of the copper-based device was attributed to its low sheet-resistance and the high work-function of copper. This could markedly reduce the potential barrier and aid efficient hole-injection into the emissive layer. The turn-on voltages were close to that of an indium tin oxide based device, but the luminescence and current efficiencies were relatively lower. For an injection-current of 20mA/cm^2, the forward voltages were 6.5, 5.8, 5.5 and 7.2V for

Materials Research Forum LLC
https://doi.org/10.21741/9781644903476

organic devices with gold, silver, copper and indium tin oxide electrodes, respectively. The changes in forward voltages were attributed to sheet-resistance variations. A stable emitting-layer peak position for all of the devices indicated that the emission was independent of nature of the electrode material. A wet-chemical lift-off process was used[152] to create a network of flexible gold transparent electrodes by using an electro-spun polymer fibre network as a mask. A resistance of 5.18Ω/sq was found for the transparent electrode while the transparency was about 90%. A root-mean-square roughness of 23nm was found when the gold nano-wire thickness was 30nm. A perovskite $CH_3NH_3PbI_3$ photo-detector which was based upon a 30nm-thick gold-network electrode had a linear dynamic range of 138dB, a detectivity of over 10^{12} Jones and a better flexibility than an existing indium tin oxide electrode.

Table 16. Characteristics of Cu-TCE and ITO-based organic light-emitting devices

Parameter	Cu-TCE	ITO
maximum luminance	$4300cd/m^2$	$4315cd/m^2$
maximum current density	$800A/m^2$	$830A/m^2$
maximum current efficiency	5.38cd/A	5.12cd/A
current efficiency(8V)	0.33cd/A	0.75cd/A
turn-on voltage	5.5V	5.7V

TCE: transparent conductive electrode, ITO: indium tin oxide

Screen-printing and flash-light sintering were explored for the large-area high-resolution patterning of silver nano-wires[153]. The flash-light sintering permitted low-temperature processing, short operational times and a high production-rate. The silver nano-wire patterns with an area of 200mm x 200mm offered a sheet-resistance of 1.1 to 9.2Ω/sq, a transparency of 75.2 to 92.6%, an efficiency of up to 200mm/s, a line-width of about 50μm, an electrical conductivity greater than 2×10^6S/m and a figure-of-merit of above 1100. Pristine nano-wires had loose contact points, and this resulted in a high electrical resistance, but flash-sintered silver nano-wires had strong welded junctions with the gaps between the nano-wires became smaller or disappeared. The welded junctions decreased the resistance of the nano-wire patterns. Under a flash-lamp at a distance of 5mm, the nano-wires were illuminated to high-intensity pulsed light with a wavelength of 200 to

1500nm. This generated highly localized heat only at the junctions. Samples were sintered using various pulse-lengths while maintaining a pulse-voltage of 350V and a pulse-number of 40. The sheet-resistance decreased essentially linearly from 46 to 11Ω/sq as the pulse-length was increased from 100 to 600μs. The resistance then remained almost constant at about 10.3Ω/sq when the pulse-length was further increased to 900μs, where the energy-density corresponded to 1.385J/cm^2. The transmittance of the silver nano-wire electrodes remained almost constant at about 91.5% during the flash-sintering when exposed to an energy-density of 1.385J/cm^2.

A compact high gain high-efficiency transparent Fabry-Perot cavity antenna was designed[154] that comprised a transparent metal mesh and which exhibited a low sensitivity to the sheet-resistance of the material. A maximum radiation efficiency of 67.1%, a peak gain of 14.2dBi and a gain band-width of 9.4% were obtained. The device consisted of dual layers of transparent PRS, a transparent patch antenna, a loaded transparent artificial magnetic conductor structure plus glass holders and substrate. Transparent metal mesh having a sheet-resistance of 0.25Ω/sq was used for the patch antenna and the bottom ground.

A printing-based low-temperature process was proposed[155] for the creation of high-resolution patterned layers with line-widths as small as 1μm. The method was based upon reverse-offset printing of a sacrificial polymer resist, followed by vacuum-deposition and lift-off. The shape of the resist layer edges permitted the patterning of evaporated aluminium, SiO and indium tin oxide. The utility of the method was initially demonstrated by using aluminium with a resistivity of about 5 x 10$^{-8}\Omega$m as a transparent metal-mesh conductor with a circa 35Ω/sq sheet-resistance and 85% area of transparency for electrodes in an In$_2$O$_3$ device. Grid line-widths ranging from 1 to 8μm and grid pitches ranging from 6.25 to 100μm could be patterned by using 40nm-thick aluminium. Grids with a constant transparency area of 0.71 could be obtained by using widths of 1, 2 or 4μm and corresponding pitches of 6.25, 12.5 or 5μm. Failures in the form of filled grid spaces or missing lines could occur; especially for the densest grids. For a transparency area of 0.85, the resistance was about 35Ωsq for any line-width. Aluminium which was patterned by using the hybrid method could be used as source/drain contacts for solution-processed In$_2$O$_3$.

Going beyond mesh, it was shown, experimentally and theoretically[156], that oblique metal gratings with optimum tilt-angles can become transparent to broadband terahertz waves under normal incidence conditions. Steel gratings were used to verify the theoretical results. The gratings consisted of 200μm x 1.95mm x 6cm steel strips with tilt-angles of 0° or 36°. The radiation was within the range of 0.2 to 1.2THz frequency range and was normalized with respect to the transmission spectrum of air. The mechanism underlying the high broadband transmission for normal incidence could be understood at the microscopic level. For a normal grating (0°) the incident electric field drove the movement of free electrons on the surface so as to form two oscillating dipoles. They eventually formed Fabry-Perot resonance peaks. Direct imaging proved the broadband transparency of structured metals, and that this transparency was insensitive to the grating thickness. This was due to the non-resonance mechanism. The optimum tilt-angle was moreover determined only by the strip-width and the grating period.

It was demonstrated that metallic gratings with narrow slits can become highly transparent to very wide bandwidths under oblique incidence[157]. Broadband optical transmission was confirmed for structured metals with a thickness within the range of half a wavelength, and the high transmission efficiency was insensitive to the metal thickness. Thick metal gratings offer the advantages of extremely high-efficiency transmission and high-grade electrical properties. Experiments were performed in the THz band, but continuous and essentially unitary transmission was possible at broadband frequencies ranging from visible light to radio. A simple means for creating broadband transparent metals was proposed. In order to confirm the transparency of oblique gratings under normal incidence, two thick gratings with lattice parameters of 0.3 and 0.56mm were prepared. One of them had an oblique angle of 32° and the other had an angle of 0°. The spectra of the gratings were very different, even though they had the same effective ratio (0.464). The horizontal period of the oblique grating was 0.66mm and the first-order Wood anomaly shifted to 0.66mm, from the 0.56mm of a normal grating. The average transmission efficiency in the long-wavelength range was much higher than that of a normal grating. The transmittance tended to be flat in the range from 0.7 to 1.0mm, with a maximum transmittance of nearly 100%. By tuning the ratio and angle of the oblique grating, a high broadband transmittance was possible for normal incidence.

Applications

Solar Cells

Carrier-selective contacts, with metal-oxide thin films, have been used in dopant-free silicon solar cells, although the properties of films such as MoO_x can soon deteriorate due to air-exposure. The use of an ultra-thin gold capping layer on the MoO_x can remedy that problem, with the gold layer also acting as a transparent conductive electrode and thus replacing transparent conductive oxides such as indium tin oxide. The power-conversion efficiency of a simple $Au/MoO_x/n\text{-}Si$ device was found[158] to increase from 0.53 to 6.43% upon using a grid-type electrode at the front surface. The gold film was investigated as both an air-exposure barrier for MoO_x and as a transparent conducting electrode. The efficiency of a silicon solar cell with a MoO_x hole-selective contact was 4.2% for a 12nm thick gold film but was negligible (0.0008%) without the film. efficiency. For these $Au/MoO_x/n$-Si solar cells, the fill-factor was the most important parameter and this was the reason for adding the metal grid-electrode.

Table 17. Performance of gold/MoO$_x$/n-Si solar cells as a function of gold thickness

Gold(nm)	$J_{sc}(mA/cm^2)$	$V_{oc}(mV)$	Fill-Factor(%)	Efficiency(%)
0	0.0096	501	16.5	0.0008
2	0.019	564	22.1	0.002
4	3.79	540	26.1	0.53
8	19.21	526	33.1	3.34
12	16.6	514	49.23	4.2

J_{sc}: short-circuit current-density, V_{oc}: open-circuit voltage

In conventional silicon photovoltaics, such an electrode is frequently used in order to extract a photo-generated current at the front surface, with minimum shadowing. Although 5nm-thick MoO_x was found to be the optimum, 10nm-thick MoO_x was used here in order to compare the solar-cell performance. Due to the improvement in the fill-factors of the $Au/MoO_x/n$-Si solar cells with front grid-electrodes, the solar-cell efficiency was likewise increased (tables 17 to 19).

Table 18. Performance of gold/MoO$_x$/n-Si solar cells as a function of MoO$_x$ thickness

$MoO_x(nm)$	$J_{sc}(mA/cm^2)$	$V_{oc}(mV)$	Fill-Factor(%)	Efficiency (%)
2	18.69	446	44.07	3.67
5*	17.55	531	49.81	4.64
5	17.56	474	38.05	3.16
10	16.61	514	49.23	4.2
15	15.93	523	50.08	4.17
20	15.55	509	50.22	3.98

*air-exposed, J_{sc}: short-circuit current-density, V_{oc}: open-circuit voltage

Table 19. Performance of gold/MoO$_x$(10nm)/n-Si
solar cells with a front-surface grid electrode

Gold(nm)	J$_{sc}$(mA/cm^2)	V$_{oc}$(mV)	Fill-Factor(%)	Efficiency(%)
2	9.97	577	43.2	2.49
4	19.26	546	61.14	6.43
8	17.64	517	61.55	5.61
12	14.5	503	71.29	5.21

J$_{sc}$: short-circuit current-density, V$_{oc}$: open-circuit voltage

In order to improve simultaneously both the optical transmittance and the electrical conductivity of transparent conductors, a new approach was proposed[159] for preparing electroplated nickel and copper micro-fibre networks and using them as highly-conductive transparent conductors in thin-film solar cells. The metal micro-fibres were created by using electro-spun polyacrylonitrile nano-fibres as templates. The large cross-sectional aspect-ratio of the metal-fibre networks markedly improved both the electrical conductivity and the optical transmittance (tables 20 and 21).

Table 20. Properties of nickel micro-fibre transparent
conductor solar cells with an effective area of 5.25cm^2

t$_e$(s)	Resistance(Ω/sq)	Transmittance(%)	V$_{oc}$(V)	J$_{sc}$(mA/cm^2)	FF(%)
20	1.968	91.8	0.671	29.85	47.4
40	0.697	87.8	0.648	30.47	52.1
60	0.412	81.6	0.681	29.44	63.3

t$_e$: electroplating time, J$_{sc}$: short-circuit current-density, V$_{oc}$: open-circuit voltage, FF: fill-factor

The copper micro-fibres led to a better figure-of-merit due to their very low electrical resistivity. Such metal micro-fibre transparent conductors were a likely alternative to the usual patterned grids which were used in flexible Cu(In,Ga)Se$_2$ thin-film solar cells because they could reduce the series resistance; an advantageous factor for large-area cells. The copper micro-fibres could be used as a ribbon via which to maintain solar-cell performance in cm-scale cells which were connected in series (table 22). Time-dependent conductivity degradation of the metal micro-fibres was negligible. As prepared bare samples were stored in a dry atmosphere at room temperature for some 9 months. The change in sheet-resistance ranged from -2.18% to +7.75% in the case of nickel micro-fibres, and from -3.45% to +7.23% in the case of copper micro-fibres. There was no change in the optical transmittance.

Table 21. Properties of copper micro-fibre transparent conductor solar cells with an effective area of 5.25cm²

$t_e(s)$	Resistance(Ω/sq)	Transmittance(%)	$V_{oc}(V)$	$J_{sc}(mA/cm^2)$	FF(%)
30	0.147	89.5	0.683	30.65	63.9
40	0.083	89.0	0.688	31.93	63.2
50	0.058	83.7	0.683	30.18	67.2

t_e: electroplating time, J_{sc}: short-circuit current-density, V_{oc}: open-circuit voltage, FF: fill-factor

Table 22. Properties of separate and serially-connected copper micro-fibre transparent conductor solar cells

State	$t_e(s)$	R(Ω/sq)	$T_a(\%)$	$V_{oc}(V)$	$J_{sc}(mA/cm^2)$	P(mW)	A(cm²)
separate	40	0.126	89.0	0.689	30.12	54.03	4.125
separate	50	0.037	83.7	0.685	29.05	69.47	5.25
connected	-	-	-	1.355	14.54	113.75	9.375

R: sheet-resistance, T_a: average transmittance, t_e: electroplating time, J_{sc}: short-circuit current-density, V_{oc}: open-circuit voltage, A: effective area, P: power

Organic solar cells

Semi-transparent organic solar cells can be used in applications such as electricity-generating windows, due to their tunable absorption spectrum and power-conversion efficiency (tables 23 and 24). It was demonstrated[160] that such cells could maintain over 85% of the efficiency of opaque organic solar cells, at an average visible transmittance of over 22.5%. This was due to highly transparent gold-grid front-electrodes having an average transmittance of over 95%. These gold grids, which were prepared via mask-less direct-write lithography, led to great tunability of the morphology, transmittance and resistance. As a result, the semi-transparent organic solar cells offered a power-conversion efficiency of 10.14%; corresponding to an opaque-device efficiency of 11.93%.

Table 23. Gold-grid transparent electrodes

Pitch(μm)	Sheet Resistance(Ω/sq)	AVT(%).)
20	6.14	8.9
30	8.5	61.3
50	13.2	75.1
100	23	85.7
200	41	92.4
400	74	95.7

AVT: average visible transmittance at 350-800nm

The efficiency was expected to exceed 16% at average visible transmittances above 20%. The light-utilization efficiency is one of the most important figures-of-merit, and decreased from 3.09 to 1.02 with increasing silver thickness. In these semi-transparent organic solar cells, in order to obtain a high power-conversion efficiency at an average visible transmittance of above 20%, the thickness of the silver film is usually limited to below 20nm in order to ensure adequate transparency. This leads to worsened charge transport and collection. The present highly-transparent front gold grids offered an average visible transmittance of up to 95.7%. This was an improvement of 11% over that of 80nm-thick indium tin oxide film with an average visible transmittance of 84.65% and a similar conductivity. This improvement in front-mounted transparent electrodes led to a much larger flexibility of silver film thickness when trading-off transparency against sheet-resistance.

Perovskite solar cells

The main difficulty with the incorporation of transparent metal-based conducting electrodes into perovskite solar cells is a degradation which results from the interdiffusion of metal and halide ions between the metal electrode and the perovskite layer. The insertion of a layer at the metal/perovskite interface is an obvious way to prevent that.

Table 24. Properties of semi-transparent organic solar cells

Silver(nm)	J_{sc}(mA/cm^2)	V_{oc}(V)	Fill-Factor(%)	PCE(%)	AVT(%)	LUE
15	19.39	0.682	0.707	9.32	33.2	3.09
30	20.66	0.685	0.708	10.14	22.5	2.28
50	22.49	0.688	0.711	11.11	14.6	1.62
100	25.44	0.691	0.701	12.32	-	-

J_{sc}: short-circuit current-density, V_{oc}: open-circuit voltage, AVT: average visible transmittance at 350-800nm, PCE: power conversion efficiency, LUE: light utilization efficiency

Metal nanostructure-based hybrid electrodes are usually made from transparent conductive oxides such as indium tin oxide and metal-doped zinc oxide, but such layers suffer from mechanical stresses and this limits their use in flexible perovskite solar cells. Flexible hybrid transparent conductive electrodes were prepared[161] which consisted of a copper grid-embedded polyimide film and a graphene capping layer. These offered excellent mechanical and chemical stability together with good opto-electrical properties.

Table 25. Properties of perovskite solar cells with various electrodes

Electrode	V_{oc}(V)	Fill-Factor	Power Conversion Efficiency(%)	J_{sc}(mA/cm^2)
glass/ITO	22.6	0.99	0.78	17.5
PET/ITO	20.1	0.97	0.77	15.1
GCEP	21.7	0.99	0.76	16.4

GCEP: copper grid-embedded polyimide film with graphene layer, PET: polyethylene terephthalate, ITO: indium tin oxide, J_{sc}: short-circuit current-density, V_{oc}: open-circuit voltage

The graphene played a critical role as a protective layer which prevented metal-induced degradation and halide diffusion between the electrode and the perovskite layer. The behaviour of the flexible devices was comparable to that of rigid equivalents which were based upon indium tin oxide. The graphene was an effective diffusion barrier for use with metallic nanostructures because of its marked impermeability to metal and halide diffusion. It also offered increased charge collection across the spaces of the metal nanostructures. There was also a minimal loss of optical transmittance, due to its high optical transparency, and an improvement in the durability of the hybrid electrode. In order to investigate the interdiffusion of each element, an analysis was made of the distribution of copper and halide ions within the layers of devices after aging (100C, 36h). The migration of the constituent elements was found to be suppressed and to be confined to the layers. The behaviour of flexible perovskite solar cells which consisted of a graphite capping layer, poly(3,4-ethylene dioxythiophene):poly(styrenesulfonate), perovskite, phenyl-C61-butyric acid methyl ester and ZnO nano-particles was determined (table 25). A cell which was based upon a copper grid-embedded polyimide offered hardly any solar-cell performance due to corrosion of the copper by halide ions and poor charge-collection. When graphite was added, a power-conversion efficiency of 16.4% measured with an open-circuit voltage of 0.99V, a short-circuit current density of 21.7mA/cm^2, and a fill-factor of 76% in the reverse scan direction.

Table 26. Properties of perovskite solar cells for reverse and forward scan directions

Electrode	Direction	J_{sc}(mA/cm^2)	V_{oc}(V)	Fill-Factor	PCE(%)
glass/ITO	reverse	22.5	0.98	0.76	16.9
glass/ITO	forward	22.4	0.97	0.74	16.3
GCEP	reverse	21.6	0.97	0.75	15.9
GCEP	forward	21.5	0.96	0.73	15.2

GCEP: copper grid-embedded polyimide film with graphene layer, ITO: indium tin oxide, J_{sc}: short-circuit current-density, V_{oc}: open-circuit voltage, PCE: power-conversion efficiency

Reference devices with indium tin oxide substrates on glass and polyethylene terephthalate offered power-conversion efficiencies of 17.5% and 15.1%, respectively. The properties of the indium tin oxide and graphene-layer copper grid-embedded polyimide film cells were determined for forward and reverse scan directions (table 26). A useful summary of other cells with transparent metallic electrodes and various types of transparent conductive electrodes was provided (table 27).

Table 27. Properties of perovskite solar cells with transparent metal electrodes

Metal	Transmittance(%)*	R(Ω/sq)	EQE(mA/cm^2)	PCE(%)
Ag	81.0	7.5	-	7.0
Ag grid	90.0	-	19.2	14.5
Ag mesh	82-86	3	17.9	14.2
Ag mesh	66.0	19	-	16.5
Ag thin film	60.0	-	-	6.2
AgNW	80.0	18	13.5	8.4
AgNW	78.0	-	-	9.2
AgNW	88.6	11.86	-	11.2
AgNW	88.2	37	-	14.2
Au mesh	-	18.0	-	13.6
Au thin film	58.0	15.16	-	6.7
CuNW	80.0	51.4	~10	5.3
Ni thin film	45.0	63	-	8.7
Ni-Au mesh	80.0	16.9	19.3	13.9

*550nm, EQE: external quantum efficiency, PCE: power-conversion efficiency

Other Photovoltaics

It was found that an aluminium-doped ZnO layer induced the formation of epitaxial ultra-thin and ultra-smooth gold films on mica[162]. Such a film exhibited an optimum electrical conductivity, together with good optical properties and a high tolerance to bending. The sheet-resistance was 8Ω/sq and the average transmittance was 80.1%. The resistivity remained low even after 1000 continuous bending cycles. A gold/Al-doped-ZnO anode-based organic light-emitting diode was prepared which offered a maximum luminance intensity of 32540cd/cm^2. Mica was an ideal substrate for flexible opto-electronic devices because of its low cost, greater than 90% optical transmittance in the visible region and a tolerance of temperatures of up to 1000C. The coefficient of thermal expansion of about 1 x 10^{-6}/C also matched that of silicon devices. Mica was also a good van der Waals epitaxy substrate for growing 2-dimensional materials such as graphene, MoS_2, ReS_2 and inorganic thin films. When used as a layer with various thicknesses, aluminium-doped ZnO seed layers induced the formation of epitaxial ultra-thin and ultra-smooth gold film. An 8nm-thick gold film had an excellent surface morphology, with a root-mean-square roughness of 0.8nm. Organic light-emitting diodes were prepared on mica after depositing thin films of gold and aluminium-doped ZnO on (001) native muscovite mica by means of pulsed laser deposition with an energy and repetition frequency of 225mJ and 5Hz. The optimum thickness for ultra-thin metals was close to that at which metal islands started to form a continuous film and resulted in macroscopic electrical conduction. Nucleation and wettability were important for reducing the percolation thickness and increasing the continuity of the films. Because of its relatively high surface energy, the aluminium-doped ZnO seed-layer produced a favourable wetting effect for subsequent gold growth. The higher activation energy reduced the tendency to surface diffusion and mass-transport of gold. This resulted in a smooth surface topology due to the dense coalescence of smaller islands. The Volmer-Weber growth of gold also tended to be suppressed on aluminium-doped ZnO. The 8nm-thick gold film which was grown directly onto mica had a roughness of about 3.8nm, while the gold film which was grown on aluminium-doped ZnO had a roughness as small as 0.8nm. A smooth surface for the gold was obtained when the ZnO thickness was greater than 10nm. The ZnO buffer-layer also markedly reduced the percolation threshold for the formation of a continuous film: from 8nm for gold films on mica to 5nm for films on ZnO. In the case of pure gold film, an initial rapid decrease in sheet resistance with increasing thickness was followed by a very gradual decrease when the thickness was greater than 8nm. This was a natural feature of island growth. Overall, the gold/ZnO films were sufficiently conductive (80.5Ω/sq) at a thickness of 4nm to be used as a transparent electrode in opto-electronic devices, while the 4nm gold films on mica were unsuitable, with their value of $2.6 \times 10^5 \Omega$/sq. The transmittance at 550nm as a function of the thicknesses of gold on 20nm ZnO films did not simply decrease with increasing gold film thickness. This was consistent with island-growth. At an effective thickness of 8nm, gold/ZnO films could attain an average transparency of 80.1% and a sheet-resistance of about 8Ω/sq by adjusting the ZnO thickness to equal 25nm. A gold/ZnO(15nm) film exhibited a maximum percentage of transmittance, in the visible region, which attained 87.5%; together with a resistance of 17Ω/sq.

Amorphous InZnO/MoS_2 heterojunction-based phototransistors having good photo-conductive gain and responsivity over the visible range were investigated[163]. The photo-generated current of an InZnO phototransistor at wavelengths greater than 600nm was markedly improved by using narrow-bandgap MoS_2 as a capping layer. At shorter wavelengths, photocarriers were generated due to optical absorption by both the InZnO and MoS_2 layers. The latter ensured appreciable photo-carrier generation even at higher wavelengths. The photo-generated carriers then moved to the underlying InZnO layer of greater carrier mobility. This had a high current of additional electrons arising from optically-induced doubly positively charged oxygen vacancies. The dynamic photosensitivity behaviour of the phototransistor indicated the presence of a persistent photoconductivity which was due to the oxygen vacancy that was associated with InZnO. The optical properties of the photo-transistors were further improved by replacing an opaque titanium/gold electrode by an ultra-thin transparent titanium/gold electrode. Use of the transparent electrode resulted in enhanced electron injection, from source to channel, due to a lowered barrier-height under illumination. This gave rise to a ten-fold increase in the photocurrent and responsivity of the phototransistors. The increase in photocurrent was highly dependent upon position, and was greatest when the beam was placed near to the source.

The higher density-of-states and relatively narrow band-gap of multi-layer MoS_2 gives it an advantage over the monolayer material for photo-transistor purposes. The sulphide exhibited thickness-dependent band-gap properties, with the multi-layer version exhibiting indirect band-gap characteristics and thus inferior optical properties. Improved optical properties of multi-layer MoS_2 photo-transistors were obtained[164] by using see-through titanium/gold thin-film electrodes rather than conventional structures. Such see-through metal electrodes exhibited transmittances of more than 70% for 532nm visible light. This permitted incident light to reach the entire active area. The effect of contact electrodes on MoS_2 photo-transistors was investigated by comparing the present electrodes with opaque electrodes and other transparent electrodes. The transparent metal electrodes offered greatly improved optical properties.

Double-sided emission CdSe|CdZnSeS/ZnS quantum-dot light-emitting diodes with transparent nm-thick metal electrodes were produced[165] via monolayer MoS_2-mediated van der Waals epitaxy. Flexible red diodes could be obtained by spin-coating and thermal evaporation via the transfer of wafer-area MoS_2 film from sapphire to polyethylene terephthalate substrates and ZnO nanoparticle-layer surfaces. The particular van der Waals epitaxy on the MoS_2 surface greatly aided lateral film-formation for nm-thick gold films on flat polyethylene terephthalate substrates and for aluminium films on rough ZnO layered surfaces. Conductive gold and aluminium films which were grown on MoS_2 surfaces acted as anode and cathode electrodes for current injection, respectively. The, monolayer MoS_2 helped to balance the injected charges by moderating the electron injection-rate. With a 10nm-thick gold anode and a 15nm-thick aluminium cathode, the flexible devices exhibited a sub-threshold turn-on voltage of 1.8V, a maximum luminance of some $83000cd/m^2$ and a current efficiency of $31.9cd/A$. Some 36% of the total luminance arose from the aluminium cathode side emission. The luminance-current density-voltage characteristics of the devices were analyzed (table 28), with the electroluminescence emission from the polyethylene terephthalate substrate being referred to as bottom emission, while that from the transparent aluminium cathode being termed top emission. Variations in the bottom and top emission intensities were recorded sequentially. Being minority carriers in the structure, hole-injection began at about 1.5V, and this determined the threshold voltage of the bottom and top emissions at 1.7 and 1.8V. The ratio of the electroluminescence intensities from top to bottom was always between 0.55 and 0.58, and this was attributed to the greater thickness and therefore lower transmittance of the aluminium cathode. Within the operating range which was studied, the maximum luminance values for the bottom and top were about 53400 and $29600cd/m^2$, respectively, at 8.7V. Due to the better charge balance at larger biases, the maximum current efficiency values for the bottom- and top- electroluminescences were $20.7cd/A$ at $208mA/cm^2$ and $11.4cd/A$ at $166mA/cm^2$; corresponding to maximum external quantum efficiencies of 16.6 and 9.1%, respectively. Seen as an ensemble, the flexible devices attained a maximum current efficiency of $31.9cd/A$ at $208mA/cm^2$.

Table 28. Properties of double-sided emission devices

Substrate	Device	$V_T(V)$	$L_{max}(cd/m^2)_{top/bottom}$	$CE_{max}(cd/A)_{top/bottom}$	$\lambda_{EL}(nm)$
rigid	OLED	7	2523/207.6	1.98/0.1	600-650
rigid	OLED	-	1158/1173	11.6/11.7	white
rigid	OLED	~4	1900	0.8/1.8	535+670
rigid	OLED	2.5	607/2597	3.3%/11.45%	610
rigid	QLED	2.3	9450	25.5	550
rigid	QLED	2.2	41010/37050	15.2/13.7	622
flexible	QDLED	>10	-	48/54	white
flexible	QLED	1.8	29600/53400	11.4/20.7	627

QLED: quantum-dot light-emitting diode, QDLED: quantum-disk light-emitting diode, V_T: threshold voltage , L_{max}: maximum luminance , CE_{max}: maximum current efficiency, λ_{EL}: electroluminescence

Transparent Metals   Materials Research Forum LLC
Materials Research Foundations **174** (2025)   https://doi.org/10.21741/9781644903476

In transparent solar thermo-photovoltaic technology, broadband absorption at ultra-violet and infra-red frequencies and simultaneous transmission at visible frequencies is possible by creating metamaterials which include semiconducting oxides. One such optically transparent metasurface-based device[166] comprised indium tin oxide as a transparent metal and ZnS as a substrate. The structure offered an absorption rate of up to 99% in the ultra-violet region and over 90% absorption in the far-infrared range, while maintaining a high average transmittance in visible ranges. The absorption remained high and exceeded 90% even when the incident angle was less than 70° for both transverse electric and transverse magnetic polarization waves. Simulation and theoretical results indicated an absorptivity of over 90% at 250 to 400nm and 800 to 2000nm ranges.

Transparent self-powered ultra-violet photo-detectors were prepared[167] which comprised dual-asymmetry interdigitated electrodes that consisted of gold and silver of differing size on top of a ZnO active layer. The electrodes were less than 10nm thick and were very transparent in the ultra-violet and visible ranges. This rendered the entire device transparent, with an output-light current which was some 33% higher than that of similar devices with thick opaque electrodes at 0V. A photo-detector with just electrode-material asymmetry could function at 0V because of a Schottky-junction that formed at the Au/ZnO interface and the ohmic contact at a silver/ZnO interface. The junction-asymmetry was additionally increased by an electrode size-difference. A photo-detector with a gold to silver finger-width ratio of 1:4 offered a photocurrent which was 11 times greater than that of a device having identical gold and silver sizes. A responsivity of 56.3μA/W and a detectivity of 1.54 x 10^8Jones were measured (table 29). The rise and fall times could be as low as 3.1 and 2.8ms, respectively. Upon patterning the pads, the photo-detector became more uniform and transparent, with an average visible transmissivity of 77.6%, while the opto-electronic conversion behaviour was unchanged. The overall properties could be better than those of previous similar devices.

*Table 29. Comparison of the characteristics of transparent
self-powered ultra-violet photo-detector chips*

Junction	Construction	Transmission	Response	Detection(Jones)
homo	ZnO nanofibres	89%	1mA/W	-
hetero	TiO$_2$/CuI	70%	0.3mA/W	8.40×10^{11}
hetero	NiO/ZnO/ITO	74.8%	20μA/W	7.20×10^{11}
hetero	ZnO/NiO nanofibres	90%	0.415mA/W	-
hetero	FTO/TiO$_2$/NiO/Ag NW	44.2%	0.4-1μA/W	1.11×10^9
hetero	Ti$_3$C$_2$T$_x$/NiO/TiO$_2$/FTO	33%	20mA/W	4.10×10^{10}
hetero	CuI/ZnS/ZnO nanorods	>70%	43.8mA/W	3.84×10^{14}
Schottky	FTO/TiO$_2$/Ag NW	70%	32.5m/W	6.00×10^9
Schottky	ZnO MSM+Au aIDT	74.3%	0.64μ/W	1.05×10^8
Schottky	ZnO MSM+Au-Ag da-IDT,u	71.7%	56.3μa/W	1.54×10^8
Schottky	ZnO MSM+Au-Ag da-IDT,p	77.6%	41.1μA/W	1.23×10^8

NW: nanowire, ITO: indium tin oxide, FTO: fluorine-doped tin oxide, MSM: metal-semiconductor-metal, IDT: interdigitated, a: asymmetrical, da: dual-asymmetry, u: unpatterned, p: patterned

The production of flexible transparent metal electrodes for flexible organic light-emitting devices is difficult because ultra-thin metallic layers tend to exhibit island-like growth. Flexible and transparent ultra-thin silver electrodes offering a high mechanical stability and good opto-electrical properties were obtained[168] by tailoring the surface properties of plastic substrates with the use of an ultraviolet-ozone treatment. This controlled the nucleation and growth kinetics of the silver films. Composite transparent electrodes of silver(9nm)|MoO$_3$(20nm), prepared on an ultraviolet-ozone treated polyethylene terephthalate substrate offered a sheet-resistance of about 7.9Ω/sq, an optical transmittance of some 87.2% at 550nm, an environmental stability of 30 days at 65C in 80% humidity and a mechanical flexibility of 100000 bending cycles at a bending radius of 1.5mm. These properties were attributed to the surface treatment of the polyethylene terephthalate substrate using ultraviolet and ozone. This increased the substrate's surface energy and generated nucleation sites among phenolic hydroxyl groups. These groups provided efficient nucleation sites for subsequent silver-film growth and also formed C-

O-Ag bonds between the substrate surface and the silver layer. These acted as so-called anchor-chains which solidly fixed the silver atoms to the substrate surface. Ultra-thin silver composite electrodes on flexible white polyethylene terephthalate, polyethylene naphthalate or Norland adhesive substrates offered maximum current efficiencies of 53.0, 77.0 and 65.2cd/A, respectively.

By combining printing, thin-film deposition and wet-etching, interconnected transparent metal-micromesh electrodes were produced[169]. Blade-coating generated self-assembled polymer micromesh networks on flexible substrates, and the network structures were then converted into conductive metal networks. The as-prepared metal-mesh films had a surface roughness of about 20nm and a thickness of as little as 50nm. A transmittance of 86% and a conductivity of 80Ω/sq was possible. The electrodes exhibited mechanical flexibility, with no conductivity degradation occurring at the smallest bending radius (1mm) after repeated bending for 3000 cycles at a bend-radius of 15mm. Organic light-emitting diodes were produced by using transparent metal-mesh electrodes and the blade-coating technique. In this process, a polystyrene-sphere suspension in water was firstly blade-coated onto a flexible poly(ethylene terephthalate) or poly(ethylene naphthalate) substrate. The printed diodes had a turn-on voltage of 3.4V and could attain a luminance of over 4000cd/m^2 at 6.5V. At a luminance of 100cd/m^2, the diodes exhibited a current density of 7.6mA/cm^2, an external quantum efficiency of 3.6% and a luminous efficiency of 1.4lm/W. The organic layer was usually very thin for opto-electronic devices and this hindered the utilization of many transparent metal-mesh electrodes having high conductivity and transparency but high roughness. Low roughness is a critical factor for the conductors in thin-film devices. Due to the flexibility and appreciable static electricity of the plastic, it was difficult to make large-area surface morphology measurements via atomic force microscopy. The structures and properties of the electrodes could be tuned by varying the polystyrene particle suspension concentration from 35 to 65mg/ml. Lower solution concentrations led to micromesh networks with narrower widths and larger spaces. Upon decreasing the suspension concentration, the electrode's optical transmittance gradually increased while its electrical conductivity decreased. At a solution concentration of 35mg/ml, a disconnection the micromesh network branches led to a marked decrease in the conductivity. More locations were disconnected at lower concentrations, and this led to non-conduction of the networks. Thinner films led to a slightly better transmittance, with a loss in conductivity. Bare-metal film, with 6nm of chromium and 9nm of gold, had a resistance of 766kΩ/sq. This did not satisfy the requirements for a conductive electrode. For a given blade-coating solution concentration, an increase in the metal thickness reduced the sheet resistance without significantly affecting the transmission.

Transparent Metals
Materials Research Foundations **174** (2025)

Materials Research Forum LLC
https://doi.org/10.21741/9781644903476

Heaters

Electro-active shape-memory polymer composites offer a poor electric-heating efficiency, and the transparency of the matrix is greatly impaired by the presence of the conductive fillers. Transparent metal mesh has therefore been used[170] to replace the conductive fillers when constructing a conductive network. Self-cracking aluminium mesh was embedded in transparent shape-memory polyimide film by means of solution-casting. The cracking was produced by using a template that consisted mainly of a large number of quadrilateral micro-cracks of similar geometry. The morphology of the micro-cracks was such that their width ranged from 2 to 6μm, with an average value of 4.0μm. The resultant flexible transparent heater offered a rapid response and a high steady-state temperature. The transmittance at 550nm was 83% and the sheet-resistance was 3.0Ω/sq. It reached a steady-state temperature of 235C within 20s. As an electrical actuator, the material could be active and deformable, depending upon its stiffness setting, and reverted to its original shape within 13s under electric stimulation. The transition temperature of 230C was much higher than that of rival electro-active shape-memory polymer composites. The present material was attached to transparent shape-memory polystyrene as a flexible transparent heater, and deformation was triggered by applying an electric field.

Diffraction Grating

A key component of certain display systems is a see-through reflective grating. For a reflective meta-surface, transmission is usually impossible due to a non-transparent metal back-plate. A method was proposed[171] which created see-through metal-dielectric meta-surfaces by etching apertures, of random position and diameter, which were much larger than the wavelength of the meta-surfaces. A 1200lp/mm metal-dielectric meta-surface diffraction grating was designed for use in the reflection of 650nm illumination. The device offered a 20% diffraction efficiency for angles of incidence of 0 to 50°. It was semi-transparent and allowed some 50% of the light to illuminate the back of the device via the random apertures.

Antennae

An optically transparent magneto-electric dipole antenna with a radiation pattern covering a wide impedance bandwidth was constructed[172] by combining a flexible transparent metal-mesh film with a low-loss quartz glass so as to form a transparent 3-dimensional radiating structure. When compared with conventional 2-dimensional methods, the present construction method provided improved radiation without losing compactness. An antenna element with a relative impedance bandwidth of some 42.3% and a peak gain of 4.8dBi had an optical transparency (400-800nm) of 51 to 82%. The lower limit was caused by the overlap where 3 stacked metal-mesh film layers coincided. A transparent metal-mesh film generally consists of two layers: a transparent substrate which acts as a base layer and an ultra-thin (<10μm) patterned metal layer; the metal-mesh layer atop it. The metal-mesh layer can have various grid patterns: square, diamond, equilateral hexagon, serpentine grid. A square grid was chosen here. The film's transparency depends upon the optical transparency of the metal-mesh conductive layer and that of the substrate. By inserting thin metal mesh film through slots in thick glass it was possible to make a vertical electrical connection by direct contact. This made it possible to create a 3-dimensional transparent antenna by creating so-called invisible conducting vias rather than 2-dimensional planar structures.

A large-scale high-gain high-efficiency optically transparent grid-array antenna which operated at mm frequencies was developed[173]. The grid was used to reduce the feeding-network losses which were caused by a lossy transparent conductor, by using a series of rectangular loops which were directly connected to their neighbours rather than being connected via a network. By loading rectangular patches onto the radiators of the outer region, the radiation and impedance behaviours were improved. A large-scale array (5.4λ x 5.1λ x 0.04λ) for 25GHz was created by using a transparent metal mesh which had a sheet resistance of 0.25Ω/sq. A peak gain of 18.18dBi, a radiation efficiency of 59.6% and an optical transparency of 84% were thus obtained (table 30).

Table 30. Properties of transparent antennae

Type	Frequency(GHz)	Gain(dBi)	RE(%)	Size(λ^3)
micro-strip array	28	9.16	41.2	3.76x3.76x0.05
reflect array	26	22.2	-	5.5x5.5x5.8
micro-strip array	56.3	9.55	-	1.3x2.06x0.04
micro-strip patch	26.2	4.83	45	1.59x1.39x0.04
reflect array	28	25.8	73	9x9x6.3
micro-strip array	26	9.7	61	3.9x1.96
micro-strip array	25	18.18	59.6	5.4x5.1x0.04
micro-strip array	25	14.64	63.5	2.83x1.33x0.04

RE: radiation efficiency

Materials Research Forum LLC
https://doi.org/10.21741/9781644903476

Supercapacitors

Metal oxides are attractive materials for the construction of supercapacitors because of their high energy-densities, but their use as transparent supercapacitors is limited by opaqueness. A further problem was the construction of flexible transparent metal-oxide supercapacitor electrodes. One solution involved $MnO_2|AuNF$ networks. A high-performance gold nano-fibre network electrode having a sheet resistance of $9.58\Omega/sq$ and an optical transparency of about 93.13% was created[174] by using an electrospinning method and thermal vacuum deposition. With the gold nano-fibre network electrode as a current-collector, an hierarchal MnO_2 nano-sheet was electrodeposited over the nano-fibre network so as to create a highly interconnected core-shell $MnO_2|AuNF$ network electrode structure with a transparency of about 86%. This $MnO_2|AuNF$ electrode had an areal capacitance of $8.26mF/cm^2$ at $5mV/s$, together with a high rate capability, a long-term cycling stability and high mechanical flexibility. The assembled flexible transparent supercapacitor had a transparency of about 79% and an energy-density of $0.14\mu Wh/cm^2$ at a power density of $4\mu W/cm^2$; together with high mechanical flexibility. The transparency and electrochemical performance of the core-shell network electrode could be tailored by controlling the electrodeposition time.

About the Author

Dr. Fisher has wide knowledge and experience of the fields of engineering, metallurgy and solid-state physics, beginning with work at Rolls-Royce Aero Engines on turbine-blade research, related to the Concord supersonic passenger-aircraft project, which led to a BSc degree (1971) from the University of Wales. This was followed by theoretical and experimental work on the directional solidification of eutectic alloys having the ultimate aim of developing composite turbine blades. This work led to a doctoral degree (1978) from the Swiss Federal Institute of Technology (Lausanne). He then acted for many years as an editor of various academic journals, in particular *Defect and Diffusion Forum*. In recent years he has specialized in writing monographs which introduce readers to the most rapidly developing ideas in the fields of engineering, metallurgy and solid-state physics. He is co-author of the widely-cited student textbook, *Fundamentals of Solidification*. Google Scholar credits him with 9614 citations and a lifetime h-index of 13.

References

[1] Jakšić Z., Maksimović M., Sarajlić M., Journal of Optics A, 7, 2004, 51. https://doi.org/10.1088/1464-4258/7/1/008

[2] Scalora M., Bloemer M.J., Manka A.S., Pethel S.D., Dowling J.P., Bowden C.M., Journal of Applied Physics, 83, 1998, 2377-2383. https://doi.org/10.1063/1.366996

[3] Sarto M.S., Sarto F., Larciprete M.C., Scalora M., D'Amore M., Sibilia C., Bertolotti M., IEEE Transactions on Electromagnetic Compatibility, 45[4] 2003, 586-594. https://doi.org/10.1109/TEMC.2003.819057

[4] Cattin L., Bernède J.C., Morsli M., Physica Status Solidi A, 210, 2013, 1047-1061. https://doi.org/10.1002/pssa.201228089

[5] Ji C., Liu D., Zhang C., Guo L.J., Nature Communications, 11, 2020, 3367. https://doi.org/10.1038/s41467-020-17107-6

[6] Sannicolo T., Lagrange M., Cabos A., Celle C., Simonato J.P., Bellet D., Small, 12, 2016, 6052-6075. https://doi.org/10.1002/smll.201602581

[7] Patil J.J., Chae W.H., Trebach A., Carter K.J., Lee E., Sannicolo T., Grossman J.C., Advanced Materials, 23, 2021, 2004356.

[8] Chae W.H., Sannicolo T., Grossman J.C., ACS Applied Materials and Interfaces, 12, 2020, 17909-17920. https://doi.org/10.1021/acsami.0c03587

[9] Wu H., Kong D., Ruan Z., Hsu P.C., Wang S., Yu Z., Carney T.J., Hu L., Fan S., Cui Y., Nature Nanotechnology, 8, 2013, 421-425. https://doi.org/10.1038/nnano.2013.84

[10] Hsu P.C., Kong D., Wang S., Wang H., Welch A.J., Wu H., Cui Y., Journal of the American Chemical Society, 136, 2014, 10593-10596. https://doi.org/10.1021/ja505741e

[11] Fazilaty M., Pourahmadi M., Shayesteh M.Z., Hashemian S., Journal of Physics and Chemistry of Solids, 148, 2021, 109683. https://doi.org/10.1016/j.jpcs.2020.109683

[12] Ibitoye A.I., Opetubo O., Oyinbo S.T., Jen T.C., Sibanda D., Oluwatoyin E., Engineered Science, 20, 2022, 364-376.

[13] Han S., Ju B.K., Yang C., Thin Solid Films 757, 2022. 139388. https://doi.org/10.1016/j.tsf.2022.139388

[14] Lee J.Y., Connor S.T., Cui Y., Peumans P., Nano Letters, 8, 2008, 689-692. https://doi.org/10.1021/nl073296g

[15] Kim T.W., Lee J.S., Kim Y.C., Joo Y.C., Kim B.J. *Materials, 12[15]* 2019, 2490. https://doi.org/10.3390/ma12152490

[16] Zhang Y., Liu Z., Ji C., Chen X., Hou G., Li Y., Zhou X., Cui X., Yang X., Ren C., ACS Applied Energy Materials, 4, 2021, 6553-6561. https://doi.org/10.1021/acsaem.1c00586

[17] Matsuda T., Tateishi Y., Yamashita K., Kitamura M., Kita T., Journal of the Society of Materials Science, Japan, 65[9] 2016, 642-646. https://doi.org/10.2472/jsms.65.642

[18] Liu Z., Zou Y., Ji C., Chen X., Hou G., Zhang C., Wan X., Guo L.J., Zhao Y., Zhang X., ACS Applied Materials and Interfaces, 13, 2021, 58539-58551. https://doi.org/10.1021/acsami.1c16691

[19] Crupi I., Boscarino S., Strano V., Mirabella S., Simone F., Terrasi A., Thin Solid Films, 520[13] 2012, 4432-4443. https://doi.org/10.1016/j.tsf.2012.02.080

[20] Gao J., Kempa K., Giersig M., Akinoglu E.M., Han B., Li R., Advances in Physics, 65[6] 2016, 553-617. https://doi.org/10.1080/00018732.2016.1226804

[21] Feng S., Elson J., Overfelt P.L., Optics Express, 13[11] 2005, 4113-4124. https://doi.org/10.1364/OPEX.13.004113

[22] Park Y.B., Lee S., Tobah M., Ma T., Guo L.J., Optical Materials Express, 13[2] 2023, 304-347. https://doi.org/10.1364/OME.473277

[23] Khan A., Lee S., Jang T., Xiong Z., Zhang C., Tang J., Guo L.J., Li W.D., Small, 12, 2016, 3021-3030. https://doi.org/10.1002/smll.201600309

[24] Rakete C., Baumbach C., Goldschmidt A., Samberg D., Schroer C.G., Breede F., Stenzel C., Zimmermann G., Pickmann C., Houltz Y., Lockowandt C., Svenonius O., Wiklund P., Mathiesen R.H., Review of Scientific Instruments, 82, 2011, 105108. https://doi.org/10.1063/1.3650468

[25] Nguyen-Thi H., Salvo L., Mathiesen R.H., Arnberg L., Billia B., Suery M., Reinhart G., Comptes Rendus - Physique, 13, 2012, 237. https://doi.org/10.1016/j.crhy.2011.11.010

[26] Mirihanage W.U., Falch K.V., Snigireva I., Snigirev A., Li Y.J., Arnberg L., , Mathiesen R.H., Acta Materialia, 81, 2014, 241. https://doi.org/10.1016/j.actamat.2014.08.016

[27] Shahani A.J., Xiao X., Lauridsen E.M., Voorhees P.W., Materials Research Letters, 8, 2020, 462. https://doi.org/10.1080/21663831.2020.1809544

[28] Neumann-Heyme, H., Shevchenko N., Grenzer J., Eckert K., Beckermann C., Eckert S., Physical Review Materials, 6, 2022, 063401. https://doi.org/10.1103/PhysRevMaterials.6.063401

[29] Gibbs J.W., Mohan K.A., Gulsoy E.B., Shahani A.J., Xiao X., Bouman C.A., De Graef M., Voorhees P.W., Nature – Scientific Reports, 5, 2015, 11824. https://doi.org/10.1038/srep11824

[30] Zhang C., Ji C., Park Y.B., Guo L.J., Advanced Optical Materials, 9, 2020, 2001298. https://doi.org/10.1002/adom.202001298

[31] Cho C., Kang P., Taqieddin A., Jing, Y., Yong K., Kim J.M., Haque M.F., Aluru N.R., Nam S.W., Nature Electronics, 4, 2021, 126-133. https://doi.org/10.1038/s41928-021-00538-4

[32] Fahland M., Vogt T., Schoenberger W., Schiller N., Thin Solid Films, 516[17] 2008, 5777-5780. https://doi.org/10.1016/j.tsf.2007.10.032

[33] Ghosh D., Martinez L., Giurgola S., Vergani P., Pruneri V., Optics Letters, 34, 2009, 325-327. https://doi.org/10.1364/OL.34.000325

[34] Jang S., Jung W.B., Kim C., Won P., Lee S.G., Cho K.M., Jin M.L., An C.J., Jeon H.J., Ko S.H., Kim, T.S., Jung H.T., Nanoscale, 8, 2016, 14257-14263. https://doi.org/10.1039/C6NR03060B

[35] Yang C., Merlo J.M., Kong J., Xian Z., Han B., Zhou G., Gao J., Burns M.J., Kempa K., Naughton M.J., Physica Status Solidi, 215, 2018, 1700504. https://doi.org/10.1002/pssa.201700504

[36] Chen Z., Wang Z., Wang J., Chen S., Zhang B., Li Y., Yuan L., Duan Y., Nanomaterials, 13, 2023, 25. https://doi.org/10.3390/nano13010025

[37] Chen W., Thoreson M.D., Ishii S., Kildishev A.V., Shalaev V.M., Optics Express, 18[5] 2010, 5124-5134. https://doi.org/10.1364/OE.18.005124

[38] He X., Cao Q., Pan J., Yang L., He S., RSC Advances, 11, 2021, 11481-11489. https://doi.org/10.1039/D1RA00549A

[39] Zou J., Li C.Z., Chang C.Y., Yip H.L., Jen A. K.Y., Advanced Materials, 26, 2014, 3618-3623. https://doi.org/10.1002/adma.201306212

[40] Huang J., Yang L., He S., Micromachines, 14, 2023, 1447. https://doi.org/10.3390/mi14071447

[41] Guillén C., Herrero J., Thin Solid Films, 520[1] 2011, 1-17. https://doi.org/10.1016/j.tsf.2011.06.091

[42] Schubert S., Meiss J., Müller-Meskamp L., Leo K., Advanced Energy Materials, 3[4] 2013, 438-443. https://doi.org/10.1002/aenm.201200903

[43] Zhang Y., Guo X., Huang J., Ren Z., Hu H., Li P., Lu X., Wu Z., Xiao T., Zhu Y., Li G., Zheng Z., NPJ Flexible Electronics, 6, 2022, 4. https://doi.org/10.1038/s41528-022-00134-2

[44] Kang H., Jung S., Jeong S., Kim G., Lee K., Nature Communictions, 6, 2015, 6503. https://doi.org/10.1038/ncomms7503

[45] Kang H., Jung S., Jeong S., Kim G., Lee K., ACS Applied Materials and Interfaces, 10, 2018, 27510-27520. https://doi.org/10.1021/acsami.8b08578

[46] Cao W., Zheng Y., Li Z., Wrzesniewski E., Hammond W.T., Xue J., Organic Electronics, 13, 2012, 2221-2228. https://doi.org/10.1016/j.orgel.2012.05.047

[47] Deng B., Hsu P.C., Chen G., Chandrashekar B.N., Liao L., Ayitimuda Z., Wu J., Guo Y., Lin L., Zhou Y., Aisijiang M., Xie Q., Cui Y., Liu Z., Peng H., Nano Letters, 15, 2015, 4206-4213. https://doi.org/10.1021/acs.nanolett.5b01531

[48] Xu X., Liu Z., He P., Yang J., Journal of Physics D, 52[6] 2019, 455401. https://doi.org/10.1088/1361-6463/ab3869

[49] Hwang B., Park M., Kima T., Han S.M., RSC Advances, 6, 2016, 67389-67395. https://doi.org/10.1039/C6RA10338C

[50] Wang D., Hauptmann J., May C., Hofstetter Y.J., Vaynzof Y., Müller T., Flexible and Printed Electronics, 6, 2021, 035001. https://doi.org/10.1088/2058-8585/abf159

[51] Lozanova V., Lalova A., Soserov L., Todorov R., Journal of Physics - Conference Series, 514, 2014, 012003. https://doi.org/10.1088/1742-6596/514/1/012003

[52] Lim D.C., Jeong J.H., Hong K., Nho S., Lee J.Y., Hoang Q.V., Lee S.K., Pyo K., Lee D., Cho S., Progress in Photovoltaics, 25, 2018, 188-195. https://doi.org/10.1002/pip.2965

[53] Lee Y., Gupta B., Tan H.H., Jagadish C., Oh J., Karuturi S., European Physics Journal - Special Topics, 231, 2022, 2933-2939. https://doi.org/10.1140/epjs/s11734-022-00544-3

[54] Tanaka H., Taniguchi M., Japanese Journal of Applied Physics, 56, 2017, 058001. https://doi.org/10.7567/JJAP.56.058001

[55] Jeong E., Choi E.A., Ikoma Y., Yu S.M., Bae J.S., Lee S.G., Han S.Z., Lee G.H., Yun J., Acta Materialia, 202, 2021, 277-289. https://doi.org/10.1016/j.actamat.2020.10.063

[56] Mondal I., Ganesha M.K., Singh A.K., Kulkarni G.U., Materials Letters, 312, 2022, 131724. https://doi.org/10.1016/j.matlet.2022.131724

[57] Lee G.W., Sim S.B., Park L.S., Nam S.Y., Molecular Crystals and Liquid Crystals, 651[1] 2017, 163-169. https://doi.org/10.1080/15421406.2017.1338076

[58] Ślusarz A.M., Komorowska K., Baraniecki T., Zelewski S.J., Kudrawiec R., ACS Sustainable Chemistry and Engineering, 10[25] 2022, 8196-8205. https://doi.org/10.1021/acssuschemeng.2c01835

[59] Song S.M., Cho S.M., Materials Chemistry and Physics, 303, 2023, 127821. https://doi.org/10.1016/j.matchemphys.2023.127821

[60] Hokari R., Kurihara K., Takada N., Hiroshima H., Applied Physics Letters, 111[6] 2017, 063107. https://doi.org/10.1063/1.4997927

[61] Wisser F.M., Eckhardt K., Nickel W., Böhlmann W., Kaskel S., Grothe J., Materials Research Bulletin, 98, 2018, 231-234. https://doi.org/10.1016/j.materresbull.2017.10.021

[62] Moon Y.G., Koo J.B., Park N.M., Oh J.Y., Na B.S., Lee S.S., Ahn S.D., Park C.W., IEEE Transactions on Electron Devices, 64[12] 2017, 5157-5162. https://doi.org/10.1109/TED.2017.2758784

[63] Hengge M., Livanov K., Zamoshchik N., Hermerschmidt F., List-Kratochvil, A.J.W., Flexible and Printed Electronics, 6, 2021, 015009. https://doi.org/10.1088/2058-8585/abe604

[64] Schneider J., Rohner P., Thureja D., Schmid M., Galliker P., Poulikakos D., Advanced Functional Materials, 26, 2016, 833-840. https://doi.org/10.1002/adfm.201503705

[65] Jang Y., Kim J., Byun D., Journal of Physics D, 46, 2013, 155103. https://doi.org/10.1088/0022-3727/46/15/155103

[66] Kiruthika S., Rao K.D.M., Kumar A., Gupta R., Kulkarni G.U., Materials Research Express, 1, 2014, 026301. https://doi.org/10.1088/2053-1591/1/2/026301

[67] Hunger C., Rao K.D.M., Gupta R., Singh C.R., Kulkarni G.U., Thelakkat M., Energy Technology, 3, 2015, 638-645. https://doi.org/10.1002/ente.201500014

[68] Haacke G., Journal of Applied Physics, 47, 1976, 4086-4089. https://doi.org/10.1063/1.323240

[69] Azulai D., Belenkova T., Gilon H., Barkay Z., Markovich G., Nano Letters. 9, 2009, 4246-4249. https://doi.org/10.1021/nl902458j

[70] Lim J.W., Lee Y.T., Pandey R., Yoo T.H., Sang B.I., Ju B.K., Hwang D.K., Choi W.K., Optics Express, 22, 2014, 26891-26899. https://doi.org/10.1364/OE.22.026891

[71] Malureanu R., Zalkovskij M., Song Z.Y., Gritti C., Andryieuski A., He Q., Zhou L., Jepsen P.U., Lavrinenko, A.V., Optics Express, 20, 2012, 22770-22782. https://doi.org/10.1364/OE.20.022770

[72] Kim K.Y., Lee B., Physical Review A, 77, 2008, 023822. https://doi.org/10.1103/PhysRevD.77.024012

[73] Allen T.W., DeCorby R.G., Optics Express, 20[S5], 2012, A579. https://doi.org/10.1364/OE.20.00A578

[74] Hooper I.R., Preist T.W., Sambles J.R., Physical Review Letters, 97, 2006, 053902. https://doi.org/10.1103/PhysRevLett.97.053902

[75] Song Z., He Q., Xiao S., Zhou L., Applied Physics Letters, 101, 2012, 181110. https://doi.org/10.1063/1.4764945

[76] Song Z., Zhang B., Optics Express, 22[6] 2014, 6519-6525. https://doi.org/10.1364/OE.22.006519

[77] Allen T.W., DeCorby R.G., Optics Express, 20[S5] 2012, A578-A588. https://doi.org/10.1364/OE.20.00A578

[78] Foss C.A., Hornyak G.L., Stockert J.A., Martin C.R., Office of Naval Research, Technical Report, 1992, No. 74.

[79] Scalora M., D'Aguanno G., Mattiucci N., Bloemer M.J., Bowden C.M., De Ceglia D., Centini M., Mandatori A., Sibilia C., Akozbek N., Cappeddu M.G., Fowler M., Haus J.W., Optics Express, 15[2] 2007, 508-523. https://doi.org/10.1364/OE.15.000508

[80] Kock W., Bell Labs Technical Journal, 27, 1948, 58. https://doi.org/10.1002/j.1538-7305.1948.tb01331.x

[81] Song Z., Xiao S., He Q., Sun S., Zhou L., Terahertz Science and Technology, 6[2] 2013, 125-145.

[82] Li Q.L., Cheung S.W., Wu D., Yuk T.I., IEEE Antennas and Propagation Letters, 16, 2017, 920-923. https://doi.org/10.1109/LAWP.2016.2614577

[83] Sarto F., Larciprete M.C., Sarto M.S., Sibilia C., Reviews on Advanced Materials Science, 5, 2003, 329-336.

[84] Jarvis B.C., Gilleland C.L., Renfro T., Gutierrez J., Parikh K., Glosser R., Landon P.B., 2005, Proceedings of SPIE, 5924, 2005, 592414. https://doi.org/10.1117/12.616871

[85] Lee S.M., Oh S., Chang S.T., ACS Applied Materials and Interfaces, 11, 2019, 4541-4550. https://doi.org/10.1021/acsami.8b17415

[86] Walia S., Gupta R., Kulkarni G.U., Energy Technology, 3, 2015, 359-365. https://doi.org/10.1002/ente.201402204

[87] Jimenéz-Vivanco M.R., García G., Carrillo J., Morales-Morales F., Coyopol A., Gracia M., Doti R., Faubert J., Lugo J.E., Nanomaterials, 10, 2020, 222. https://doi.org/10.3390/nano10020222

[88] He X., Yang L., He S., Optics Letters, 46, 2021, 4666-4669. https://doi.org/10.1109/CAC53003.2021.9727853

[89] Li W.D., Chou S.Y., Optics Express, 18[2] 2010, 931-937. https://doi.org/10.1364/OE.18.000931

[90] Huang M., Li X., Luo J., Optics Express, 28[22] 2020, 33263. https://doi.org/10.1364/OE.408872

[91] Peng G., Ke P.X., Tseng L.C., Yang C.F., Chen H.C., Photonics, 10, 2023, 804. https://doi.org/10.3390/photonics10070804

[92] Peng G., Li W.Z., Tseng L.C., Yang C.F., Nanomaterials, 13, 2023, 766. https://doi.org/10.3390/nano13040766

[93] Becerril-Castro I.B., Turino M., Pazos-Perez N., Xiaofei X., Levato T., Maier S.A., Alvarez-Puebla R.A., Giannini V., Advanced Optical Materials, 2024, 2400.

[94] Palmer S.J., Xiao X., Pazos-Perez N., Guerrini L., Correa-Duarte M.A., Maier S.A., Craster R.V., Alvarez-Puebla R.A., Giannini V., Nature Communications, 10, 2019, 2118. https://doi.org/10.1038/s41467-019-09939-8

[95] Singh H.J., Misatziou D., Wheeler C., Buendía A., Giannini V., Sánchez-Gil J.A., Werts M.H.V., Brown T., El-Sagheer A.H., Kanaras A.G., Muskens O.L., ACS Applied Optical Materials, 1, 2023, 69-77. https://doi.org/10.1021/acsaom.2c00008

[96] Xiao X., Turino M., Becerril-Castro I.B., Maier S.A., Alvarez-Puebla R.A., Giannini V., Advanced Photonics Research, 3, 2022, 2200190. https://doi.org/10.1002/adpr.202200190

[97] Putzke C., Guo C., Plisson V., Kroner M., Chervy T., Simoni M., Wevers P., Bachmann M.D., Cooper J.R., Carrington A., Kikugawa N., Fowlie J., Gariglio S., Mackenzie A.P., Burch K.S., Îmamoğlu A., Moll P.J.W., Nature Communications, 14, 2023, 3147. https://doi.org/10.1038/s41467-023-38848-0

[98] Zhang L., Zhou Y., Guo L., Zhao W., Barnes A., Zhang H.T., Eaton C., Zheng Y., Brahlek M., Haneef H.F., Podraza N.J., Chan M.H.W., Gopalan V., Rabe K.M., Engel-Herbert R., Nature Materials, 15, 2016, 204-210. https://doi.org/10.1038/nmat4493

[99] Salifu S., Olubambi P.A., Journal of the Korean Ceramic Society, 60, 2023, 24-40. https://doi.org/10.1007/s43207-022-00266-1

[100] Yuan S., Fan Z., Wang G., Chai Z., Wang T., Zhao D., Busnaina A.A., Lu X., Advanced Science, 10, 2023, 2304990. https://doi.org/10.1002/advs.202304990

[101] Chauvin A., Heu W.T.C., Buh J., Tessier P.Y., Mel A.A.E., NPJ Flexible Electronics, 3, 2019, 5. https://doi.org/10.1038/s41528-019-0049-1

[102] Wu S.H., Cossio G., Braun B., Camille F., Wu M., Yu E.T., Advanced Optical Materials, 11[6] 2023, 2202409. https://doi.org/10.1002/adom.202202409

[103] Plyusnin P.E., Shubin Y.V., Korenev S.V., Journal of Structural Chemistry, 63[3] 2022, 353-377. https://doi.org/10.1134/S0022476622030040

[104] Wang Y., Du Y., Guo D., Qiang R., Tian C., Xu P., Han X., Journal of Materials Science, 52[8] 2017, 4399-4411. https://doi.org/10.1007/s10853-016-0687-9

[105] Meyyathal P.R., Santhiya N., Umadevi S., Michelraj S., Ganesh V., Colloids and Surfaces A, 575, 2019, 237-244. https://doi.org/10.1016/j.colsurfa.2019.05.020

[106] Ryzhonkov, D.I., Levina, V.V., Dzidziguri, E.L., Khrustov, E.N., Russian Journal of Non-Ferrous Metals, 49[4] 2008, 308-313. https://doi.org/10.3103/S1067821208040184

[107] Sarkar, J., Ramanath, G., John, V., Bose, A., Advances in Polymer Science, 218[1] 2008, 235-249. https://doi.org/10.1007/12_2008_167

[108] Zhang, G., Ma, Y., Liu, Z., Fu, X., Niu, X., Qu, F., Si, C., Zheng, Y., Langmuir, 36[51] 2020, 15610-15617. https://doi.org/10.1021/acs.langmuir.0c03142

[109] Seetharamaiah, N., Seetharamaiah, N., Pathappa, N., Melo, J.S., Gurukar, S.S., Sensors and Actuators B, 245, 2017, 726-740. https://doi.org/10.1016/j.snb.2017.02.003

[110] Huang, Z., Liu, Y., Zhang, Q., Chang, X., Li, A., Deng, L., Yi, C., Yang, Y., Khashab, N.M., Gong, J., Nie, Z., Nature Communications, 7, 2016, 12147. https://doi.org/10.1038/ncomms12147

[111] Chen, T.H., Yu, C.J., Tseng, W.L., Nanoscale, 6[3] 2014, 1347-1353. https://doi.org/10.1039/C3NR04991D

[112] Bauer, P., Mougin, K., Vignal, V., Krawiec, H., Rajab, M., Buch, A., Ponthiaux, P., Faye, D., Annales de Chimie: Science des Materiaux, 40[1-2] 2016, 43-50. https://doi.org/10.3166/acsm.40.43-50

[113] Rajab, M., Mougin, K., Derivaz, M., Josien, L., Luchnikov, V., Toufaily, J., Hariri, K., Hamieh, T., Lohmus, R., Haidara, H., Colloids and Surfaces A, 484, 2015, 508-517. https://doi.org/10.1016/j.colsurfa.2015.08.035

[114]Barreca, D., Gasparotto, A., Tondello, E., Journal of Nanoscience and Nanotechnology, 5[6] 2005, 994-998. https://doi.org/10.1166/jnn.2005.130

[115] Herderick, E.D., Tresback, J.S., Vasiliev, A.L., Padture, N.P., Nanotechnology, 18[15] 2007, 155204. https://doi.org/10.1088/0957-4484/18/15/155204

[116] Yang, H., Du, M., Odoom-Wubah, T., Wang, J., Sun, D., Huang, J., Li, Q., Journal of Chemical Technology and Biotechnology, 89[9] 2014, 1410-1418. https://doi.org/10.1002/jctb.4225

[117] Johnson, C.J., Dujardin, E., Davis, S.A., Murphy, C.J., Mann, S., Journal of Materials Chemistry, 12[6] 2002, 1765-1770. https://doi.org/10.1039/b200953f

[118] Kim, J., Myung, N.V., Hur, H.G., Chemical Communications, 46[24] 2010, 4366-4368. https://doi.org/10.1039/c0cc00408a

[119] Feng, J.J., Lin, X.X., Chen, S.S., Huang, H., Wang, A.J., Sensors and Actuators, B, 247, 2017, 490-497. https://doi.org/10.1016/j.snb.2017.03.053

[120] Liu, L., Chen, L.X., Wang, A.J., Yuan, J., Shen, L., Feng, J.J., International Journal of Hydrogen Energy, 41[21] 2016, 8871-8880. https://doi.org/10.1016/j.ijhydene.2016.03.208

[121] Fu, L., Liu, K., Lyu, Z., Sun, Y., Cai, J., Wang, S., Wang, Q., Xie, S., Journal of Colloid and Interface Science, 634, 2023, 827-835. https://doi.org/10.1016/j.jcis.2022.12.091

[122] Venu, R., Ramulu, T.S., Anandakumar, S., Rani, V.S., Kim, C.G., Colloids and Surfaces A, 384[1-3] 2011, 733-738. https://doi.org/10.1016/j.colsurfa.2011.05.045

[123] Zhang, W., Dong, Q., Lu, H., Hu, B., Xie, Y., Yu, G., Journal of Alloys and Compounds, 727, 2017, 475-483. https://doi.org/10.1016/j.jallcom.2017.06.205

[124] Chou, S.W., Shyue, J.J., Chien, C.H., Chen, C.C., Chen, Y.Y., Chou, P.T., Chemistry of Materials, 24[13] 2012, 2527-2533. https://doi.org/10.1021/cm301039a

[125] Mo, J.Q., Hou, J.W., Lü, X.Y., Optoelectronics Letters, 11[6] 2015, 401-404. https://doi.org/10.1007/s11801-015-5158-z

[126] Zhang, B., Xu, P., Xie, X., Wei, H., Li, Z., MacK, N.H., Han, X., Xu, H., Wang, H.L., Journal of Materials Chemistry, 21[8] 2011, 2495-2501. https://doi.org/10.1039/C0JM02837A

[127] Rosli, M.M., Aziz, T.H.T.A., Umar, M.I.A., Nurdin, M., Umar, A.A., Journal of Electronic Materials, 51[9] 2022, 5150-5158. https://doi.org/10.1007/s11664-022-09762-w

[128] Ishaq A., Shehla H., Ali N.Z., Akram W., Shakeel K., Diallo A., Shahzad N., Maaza M., Materials Research Express, 4[7] 2017, 075055. https://doi.org/10.1088/2053-1591/aa7e60

[129] Hu, G., Zhang, W., Qiao, X., Wu, K., Chen, Q., Cai, Y., Zhang, W., Physica E, 64, 2014, 211-217. https://doi.org/10.1016/j.physe.2014.07.029

[130] Wang, Y., Zhang, Q., Wang, T., Zhou, J., Rare Metal Materials and Engineering, 40[12] 2011, 2207-2211.

[131] Yang, J., Lu, L., Wang, H., Shi, W., Zhang, H., Crystal Growth and Design, 6[9] 2006, 2155-2158. https://doi.org/10.1021/cg060143i

[132] Yadav, A., Follink, B., Funston, A.M., Chemistry of Materials, 34[19] 2022, 8987-8998. https://doi.org/10.1021/acs.chemmater.2c02494

[133] Chen Z., Boyajian N., Lin Z., Yin R.T., Obaid S.N., Tian J., Brennan J.A., Chen S.W., Miniovich A.N., Lin, L., Qi Y., Liu X., Efimov I.R., Lu L., Advanced Materials Technologies, 6[7] 2021, 2100225. https://doi.org/10.1002/admt.202100225

[134] Kamijo T., De Winter S., Panditha P., Meulenkamp E., ACS Applied Electronic Materials, 2022, 4, 698-706. https://doi.org/10.1021/acsaelm.1c01116

[135] Huang Y.J., Chen Y.F., Hsiao P.H., Lam T.N., Ko W.C., Luo M.Y., Chuang W.T., Su C.J., Chang J.H., International Journal of Molecular Sciences, 22[23] 2021, 12669. https://doi.org/10.3390/ijms222312669

[136] Zhang H., Zhu X., Tai Y., Zhou J., Li H., Li Z., Wang R., Zhang J., Zhang Y., Ge W., Zhang F., Sun L., Zhang G., Lan H., International Journal of Extreme Manufacturing, 5, 2023, 032005. https://doi.org/10.1088/2631-7990/acdc66

[137] Lee J., Varagnolo S., Walker M., Hatton R.A., Advanced Functional Materials, 30, 2020, 2005959. https://doi.org/10.1002/adfm.202005959

[138] Lee H.B., Jin W.Y., Ovhal M.M., Kumar N., Kang J.W., Journal of Materials Chemistry C, 7, 2019, 1087-1110. https://doi.org/10.1039/C8TC04423F

[139] Hong S., Lee H., Lee J., Kwon J., Han S., Suh Y.D., Cho H., Shin J., Yeo J., Ko S.H., Advanced Materials, 27, 2015, 4744-51. https://doi.org/10.1002/adma.201500917

[140] Chen Y., Fu X.Y., Yue Y.Y., Zhang N., Feng J., Sun H.B., Applied Surface Science, 467-468, 2019, 104-11. https://doi.org/10.1016/j.apsusc.2018.10.093

[141] Li K., Wang H., Li H., Li Y., Jin G., Gao L., Marco M., Duan Y., 2017 Nanotechnology, 28[31] 315201. https://doi.org/10.1088/1361-6528/aa7a6a

[142] Wang X.R., Wang X.B., Ren H., Wu N.S., Wu J.W., Su W.M., Han Y.L., Xu S., Progress in Electromagnetics Research, 170, 2021, 187-197. https://doi.org/10.2528/PIER21052101

[143] Lee M.S., Lee K., Kim S.Y., Lee H., Park J., Choi K.H., Kim H.K., Kim D.G., Lee D.Y., Nam, S.W., Park J.U., Nano Letters, 13, 2013, 2814-2821. https://doi.org/10.1021/nl401070p

[144] Li M., Zuo W.W., Ricciardulli A.G., Yang Y.G., Liu Y.H., Wang Q., Wang K.L., Li G.X., Saliba M., Di Girolamo D., Abate A., Wang Z.K., Advanced Materials, 32, 2020, 2003422. https://doi.org/10.1002/adma.202003422

[145] Li Z., Li H., Chen L., Huang J., Wang W., Wang H., Li J., Fan B., Xu Q., Song W., Solar Energy, 206, 2020, 294-300. https://doi.org/10.1016/j.solener.2020.06.013

[146] Ou X.L., Feng J., Xu M., Sun H.B., Optics Letters, 42[10] 2017, 1958-1961. https://doi.org/10.1364/OL.42.001958

[147] Kao P.C., Hsieh C.J., Chen Z.H., Chen S.H., Cells, 186, 2018, 131-141. https://doi.org/10.1016/j.solmat.2018.06.031

[148] Song S., Cho S.M., Korean Journal of Chemical Engineering, 38[8] 2021, 1720-1726. https://doi.org/10.1007/s11814-021-0811-7

[149] Hong S., Yeo J., Kim G., Kim D., Lee H., Kwon J., Lee H., Lee P., Ko S.H, ACS Nano, 7, 2013, 5024-5031. https://doi.org/10.1021/nn400432z

[150] Tian J., Lin Z., Chen Z., Obaid S.N., Efimov I.R., Lu L., Photonics, 8, 2021, 220. https://doi.org/10.3390/photonics8060220

[151] Gedda M., Das D., Iyer P.K., Kulkarni G.U., Materials Research Express, 7, 2020, 054005. https://doi.org/10.1088/2053-1591/ab8d5b

[152] Yang J., Bao C., Zhu K., Yu T., Xu Q., ACS Applied Materials and Interfaces, 10, 2018, 1996-2003. https://doi.org/10.1021/acsami.7b15205

[153] Li W., Yarali E., Bakytbekov A., Anthopoulos T.D., Shamim A., Nanotechnology, 31, 2020, 395201. https://doi.org/10.1088/1361-6528/ab9c53

[154] An K., Sun P., Chen A., Electronics Letters, 60[8] 2024, 1-3. https://doi.org/10.1049/ell2.13203

[155] Sneck A., Ailas H., Gao F., Leppäniemi J., ACS Applied Materials and Interfaces, 13, 2021, 41782-41790. https://doi.org/10.1021/acsami.1c08126

[156] Fan R.H., Li J., Peng R.W., Huang X.R., Qi D.X., Xu D.H., Ren, X.P., Wang M., Applied Physics Letters, 102, 2013, 171904. https://doi.org/10.1063/1.4803467

[157] Fan R.H., Peng R.W., Huang X.R., Li J., Liu Y., Hu Q., Wang M., Zhang X., Advanced Materials, 24, 2012, 1980-1986. https://doi.org/10.1002/adma.201104483

[158] Lee Y., Gupta B., Tan H.H., Jagadish C., Oh J., Karuturi S., European Physical Journal: Special Topics, 231[15] 2022, 2933-2939. https://doi.org/10.1140/epjs/s11734-022-00544-3

[159] Cho D.H., Lee W.J., Hwang T.H., Huh J., Yoon S.S., Chung Y.D., Journal of Power Sources, 603, 2024, 234443. https://doi.org/10.1016/j.jpowsour.2024.234443

[160] Duan X., Ding Y., Wang J., Wang Z., Wang Y., Liu R., ACS Applied Energy Materials, 6[2] 2023, 617-621. https://doi.org/10.1021/acsaem.2c03796

[161] Jeong G., Koo D., Seo J., Jung S., Choi Y., Lee J., Park H., Nano Letters, 20[5] 2020, 3718-3727. https://doi.org/10.1021/acs.nanolett.0c00663

[162] Xie J., Bi Y., Ye M., Rao Z., Shu L., Lin P., Zeng X., Ke S., Applied Physics Letters, 114[8] 2019, 081902. https://doi.org/10.1063/1.5082803

[163] Das D., Park J., Ahn M., Park S., Hur J., Jeon S., Nanotechnology, 31[3] 2020, 035201. https://doi.org/10.1088/1361-6528/ab4526

[164] Park J., Das D., Ahn M., Park S., Hur J., Jeon S., Nano Convergence, 6[1] 2019, 32. https://doi.org/10.1186/s40580-019-0202-5

[165] Huang S.C., Chang C.J., Lin Y.H., Lin S.Y., Huang C.Y., ACS Applied Nano Materials, 6[23] 2023, 21688-21694. https://doi.org/10.1021/acsanm.3c03839

[166] Shafique A., Naveed M.A., Aldaghri O., Cabrera H., Ibnaouf K.H., Madkhali N., Mehmood M.Q., Physica Scripta, 99[1] 2024, 015518. https://doi.org/10.1088/1402-4896/ad1451

[167] Weng C., Gao A., Huang J., Yang L., He S., ACS Photonics, 11[2] 2024, 561-569. https://doi.org/10.1021/acsphotonics.3c01435

[168] Lu W., Wang R., Li R., Wang Y., Wang Q., Qin Y., Chen Y., Lai W., Zhang X., ACS Applied Materials and Interfaces, 14[50] 2022, 55905-55914. https://doi.org/10.1021/acsami.2c18738

[169] Zhu J., Han D., Wu X., Ting J., Du S., Arias A.C., ACS Applied Materials and Interfaces, 12[28] 2020, 31687-31695. https://doi.org/10.1021/acsami.0c07299

[170] Huang X., Zhang F., Leng J., Applied Materials Today, 21, 2020, 100797. https://doi.org/10.1016/j.apmt.2020.100797

[171] Nikolov D.K., Cheng F., Ding L., Bauer A., Nick Vamivakas A., Rolland J.P., Optical Materials Express, 9[10] 2019, 4070-4080. https://doi.org/10.1364/OME.9.004070

[172] Liu Y., Luk K.M., IEEE Transactions on Antennas and Propagation, 72[7] 2024, 5569-5577. https://doi.org/10.1109/TAP.2024.3408234

[173] An K., Sun P., Deng Y., Chen A., IEEE Antennas and Wireless Propagation Letters, 23[5] 2024, 1598-1602. https://doi.org/10.1109/LAWP.2024.3363627

Materials Research Forum LLC
https://doi.org/10.21741/9781644903476

[174] Singh S.B., Singh T.I., Kim N.H., Lee J.H., Journal of Materials Chemistry A, 7[17] 2019, 10672-10683. https://doi.org/10.1039/C9TA00778D

www.ingramcontent.com/pod-product-compliance
Lightning Source LLC
Chambersburg PA
CBHW071714210326
41597CB00017B/2483